D0386632

The Quality Secret:
The Right Way to Manage©

The Quality Secret: The Right Way to Manage©

by William E. Conway

Conway Quality, Inc.
15 Trafalgar Square
Nashua, NH 03063
1-800-359-0099

The Quality Secret: The Right Way to Manage©. Copyright © 1992 by William E. Conway. All rights reserved. Printed in the United States of America. No part of this book may be used or reproduced in any manner whatsoever without written permission except in the case of brief quotations embodied in critical articles or reviews. For information on reordering contact Conway Quality, Inc., 15 Trafalgar Square, Nashua, NH 03063, Telephone 1-800-359-0099.

International Standard Book Number 0-9631464-0-8
Library of Congress Catalog Card Number 91-76655

This book was published by
Conway Quality, Inc.
15 Trafalgar Square
Nashua, NH 03063
Tel: 1-800-359-0099
Fax: 1-603-889-0033

10 9 8 7 6 5 4 3 2

ACKNOWLEDGMENTS

I want to thank the many people who have contributed ideas to the development of the system described in this book. It started, of course, with Dr. W. Edwards Deming. One morning in March, 1979, he completely changed my way of thinking about management. Since then, three other groups have been instrumental in the evolution of The Right Way to Manage. The large number of significant contributions make it impossible for me to name individuals, but the three groups are:

1. Customers of Conway Quality, Inc.,
2. Employees of Nashua Corporation, and
3. Colleagues at Conway Quality, Inc.

A great number of people in each group have generously shared their ideas and experiences, helping to refine the system and make it more workable in the real world.

Finally, I would like to thank Lawrence C. Hornor, Senior Associate at Conway Quality, for assistance in writing, and Brock Dethier for his editorial assistance.

CONTENTS

SECTION IV—CREATING THE NEW SYSTEM

APPENDIX

Foreword

HOW BILL DID IT

by Lawrence C. Hornor

Bill Conway was different from most CEOs. In March 1979, when most American businesspeople were just beginning to grasp the extent of the Japanese economic miracle, he was able to see quickly the importance of the principles of Dr. W. Edwards Deming and to understand what "Quality" really meant. He was also very different in that he perceived a way of developing these principles into a whole new system of management, and he took the leadership in developing and implementing the new system at Nashua Corporation. This required a major change in management thinking, a whole new culture.

It takes the perspective of hindsight to understand how difficult it was for Bill Conway to implement such a major culture change. He asked a group of managers of a successful Fortune 500 company to abandon completely their accepted "business school" approach to doing business and to adopt a radically new approach espoused by a statistician in his late 70's, a man most of us had never heard of. Bill was the first CEO of a major organization in the Western world to adopt this new system of management. There were no others. There was no one to turn to for advice and counsel. The Japanese were not talking. Bill and the rest of us had to figure it out.

I was one of those managers, and I was at least as skeptical as the rest. If this new approach was so good, why weren't any other companies using it? Besides, my divisions were doing just fine (I thought). Bill always came back from his February vacation in Barbados with some new idea, but none of them had called for a completely new way of operation. We all expressed our skepticism, but Bill was adamant. In my own mind I decided to "humor the boss." I was sure that this new concept would fade away eventually.

It soon became apparent that Bill was quite serious. He started a massive education and training program for all levels of employees. Even after attending a four-day course by Dr. Deming, I still wasn't convinced that these new management ideas were significant in the "real world" of manufacturing and marketing a high-tech product. Nevertheless I got people to keep charts—soon dubbed "charts for Bill"—plotting lots of variables in the operation. Bill would come around and ask a lot of questions about the charts and seem dissatisfied when we didn't have the answers. Admittedly some of his questions led to our discovering some waste, but the whole thing was getting to be annoying. We had more important things to do than chart everything and then explain what was happening.

After a few months, we were called to a presentation by the manufacturing superintendent for the Coating Division. He described in detail how his division had stabilized a major coating operation, shifted the level, and greatly improved the control of the process, with huge savings in material and a better, more uniform product for customers. At first I said to myself that operating a big coater was quite different than what my divisions were doing. But then I began to wonder whether we could make similar progress in the development, manufacture, and sale of hard memory discs for computers. I knew by then that Bill wasn't going to let this one die, and I was also getting intrigued by the possibility that there really was a chance to do something big.

Instead of avoiding the days when Dr. Deming was around, I began clamoring for more of his time. We began seriously trying to follow his precepts, and when Bill saw we were really getting on board, he spent a lot of his personal time helping us understand the new approach to managing the operations and working with project teams.

It seemed that Bill was everywhere. He coached, cajoled, exhorted, and even threatened until it became obvious that to work at Nashua Corporation, you would have to learn and practice the "new way" of working. Most of all he questioned, questioned, questioned. He has a facility for getting to the nub of a problem and opening your mind to new solutions. I believe this facility derives from keen observation of what is going on in the work place, a lot of hard work thinking about what he has observed, and the ability to "imagineer" new approaches.

In the latter part of 1979 and during 1980 to 1983, I would estimate Bill was spending 40% of his time promoting the change to the new system of management. He was personally very visible to all the employees working on eliminating waste. With the chief executive involved to such a degree and exhibiting a detailed understanding of what could and should be accomplished, it was impossible not to join him in the effort. People also began to understand that he respected their opinions and was looking for their help and suggestions. He was never satisfied and was persistent in looking for an ever-better way—continuous improvement, he called it.

It was not always smooth. We made false starts and wasted effort, but as we proceeded, Bill was taking what worked and building it in his own mind into a management system. In the end this system was based on what Bill called the "core activity" of improving work processes through finding, quantifying, and eliminating waste.

After being very slow to embrace the new thinking, I gradually saw the power of it, and successes in using it increased my enthusiasm. Many of those successes came with Bill's detailed help and guidance, and particularly his perceptive questions. Gradually, as Bill codified the approach, we were able to do more and more on our own. In the end, we were able to build two of my divisions, the computer memory disc business and the mail order photofinishing business, into truly world class operations, beating our competition with high, consistent quality and low costs for products and services external customers wanted.

Although we did not realize it in 1979, the advantage that other corporations had gained from this new system would also hurt our company. Our office copy business, which represented about 60% of our revenues and profits at the time, was built around the international distribution of high-quality, low-cost copy machines made by Ricoh Ltd. in Japan. Because of strong indications that Ricoh would create its own international distribution network and might deny us access to its machines, we began the major effort of developing our own line of copy machines in

1978. We felt we had no choice, because we could not find a good alternate supplier.

We designed a machine similar to Ricoh's, while carefully avoiding any patent violations. Dr. Deming's arrival encouraged us; we felt his principles would help us compete effectively with the Japanese. However, in 1980 we began to realize we had a major problem. We and our suppliers were ten to twenty years behind. Purchased parts accounted for most of the cost of the machines, and the best prices we could obtain taught us a disturbing lesson. Other than a few special parts, Japanese suppliers were reluctant to sell to Nashua. We bought some parts which were made by Japanese suppliers, sold at a profit to Ricoh, repackaged and sold by Ricoh to Nashua as spare parts, and shipped to the U.S. Including freight and duty, our cost from Ricoh was still 10–25% less than the lowest prices charged by U.S. or other non-Japanese manufacturers! It appeared that the entire U.S. metal parts industry at that time was non-competitive. With hindsight, we know we should have discovered this much sooner, but from mid-1981 on our major management challenge was extricating the company successfully from the venture. The restructuring included massive write-offs in 1981 and 1982, but the company returned to profitability in 1983.

We began to see what we faced competing against companies in Japan where entire industry groups had learned to apply Dr. Deming's principles. Somehow American companies were going to have to change their way of managing.

I was delighted to discover that when Bill retired in 1983 he would continue to teach the practical application of Dr. Deming's principles in his own firm, Conway Quality, Inc.

Why hasn't the U.S. moved faster to adopt the new system of management? I believe it is a failure of leadership. There are too few Bill Conways who have the insight to see what the new system can do and the determination required to move people to change. Gradually, however, large influential companies are making the change. Their insistence that their suppliers do the same is spreading the word. It will take more time, but at some point the competitiveness of those who have adopted the new management system will make its benefits apparent. Then the new way of management will become the only acceptable mode of operation, and the U.S. can begin to regain its place as the world's most productive nation. When that happens, I believe that Dr. Deming and Bill Conway will both deserve a lot of the credit.

INTRODUCTION

Who am I to claim to know The Right Way to Manage? You may have been managing your own way for 30 years, doing just fine. What makes me so sure about my way?

I can use such a title without apology for three reasons. First, I did not invent this management system (although I have helped to expand it and apply it to a variety of organizations); second, it is not a new, faddish system, but one that has been developing for over forty years and is based on sound statistical principles; and third, success speaks for itself. If you look at some of the American companies—and almost all of the Japanese companies—that have made remarkable gains in quality, you will find that they use variations on the same basic management approach, the approach I call The Right Way to Manage.

The Western world has lost its competitive edge, Japanese companies have gobbled up market share, and the hierarchy of industrialized countries has changed radically over the past 40 years in large part because of this management approach. Since recovering from World War II, Japanese companies have been perfecting—and gaining advantage from—a system of management born in America but largely ignored by Western organizations. Only over the past decade have leading Western companies realized how far behind they are and started adopting this system. If you and your organization are serious about changing and about surviving into the next century, read this book. It can provide you with a critical first step towards managing the new way.

Because you have read this far, you obviously know that quality has become the most important attribute of modern products and services. Yet even if you have read a good deal about quality, you may not be clear about what the quality revolution really is, how it affects you, and what you can do to move with it rather than become a casualty. I have spent the past 12 years answering questions about quality for organizations worldwide and putting into practice the ideas of the most admired expert on quality in our time, Dr. W. Edwards Deming. This book is a result of that experience. It will show you how you can make the quality and productivity of your organization the equal of any in the world.

If you do not feel that your organization needs a quality overhaul, if you do not think that it is in serious need of change, you must not be paying close attention to what is happening in the world. Within the past two decades, the United States has gone from being the world's creditor to being the greatest debtor nation the planet has ever seen. In countless industries, the U.S. has dropped from leader to has-been, outclassed and outsold by Asian and European companies.

The good news is that if you realize how bad things are now, you can take actions similar to those taken by Japanese companies during their desperate days after World War II. And you can achieve similar results. You can start following The Right Way to Manage. With Americans' well-known creativity and innovations complementing working the right way, American organizations can be the best.

Managing an organization requires skills and knowledge in a variety of areas. A good leader knows how to motivate employees, how to make the product, how to create partnerships with suppliers, how to please customers, how to keep the company afloat. Yet in business schools and in the business world in general, these skills are considered separate and discrete and are taught in courses with titles like Human Resource Management, Production, Organizational Behavior, Marketing, Finance. Managers have long needed a coherent system that ties together all of their activities. The Right Way to Manage is such a system.

Organizations will survive into the next millennium only if their leaders can coax from them goods and services of the highest possible quality. To ensure survival, these leaders must continuously strive, by eliminating waste, to improve their organizations' work processes. This book explains why and how to make such improvements.

The first section sketches a brief history of this system of management and provides working definitions of some key terms. Its examples should convince skeptical readers that this customer-driven system can work anywhere. The second section introduces the tools needed to work in the new way and demonstrates how anyone can use these tools to identify, quantify, and eliminate the waste in work processes. Section III details how the new system of management necessitates changes in the way people relate to each other, both within the organization and between organizations. The final section provides tips on how to begin the management transformation.

THE MODEL

The diagram on the following page provides a graphic overview of the concepts covered in this book. The goal of The Right Way to Manage, as the heading indicates, is to improve all work and thus please and satisfy external customers with high-quality, low-cost products and services. The core activity leading to that goal, highlighted in the red central circle, is to identify, quantify, and eliminate waste through continuously improving work processes. Chapter 8 is devoted to the pursuit of this core activity, and every part of this book relates to it.

The four ovals closest to the central circle indicate the four areas an organization must focus on in order to eliminate waste and improve quality. Variation in a process (Chapter 4) leads to variation in products and variable quality, but by studying the variation, using the tools of statistics, data gathering, and simple charts (Chapter 5), people can locate the waste and eliminate it. To use the knowledge gathered by the study of variation, people must understand as much as possible about the organization's work processes (Chapter 3). They must identify what processes actually add value to the organization's products or services, and they must reduce or eliminate the processes that do not. To improve work processes requires creativity and innovation, which can be productively channelled through the imaginative, fact-based act of imagineering (Chapter 6).

Even if it uses statistical methods to find its waste and imagineering and work improvement techniques to change its work processes, an organization cannot improve quality without also changing the way it deals with its employees, customers, and suppliers (Section III). Key to the organization's human relations are leadership, communication, and teamwork. For the new system of management to work, top leaders must embrace and lead it, and everyone must learn to work cooperatively rather than

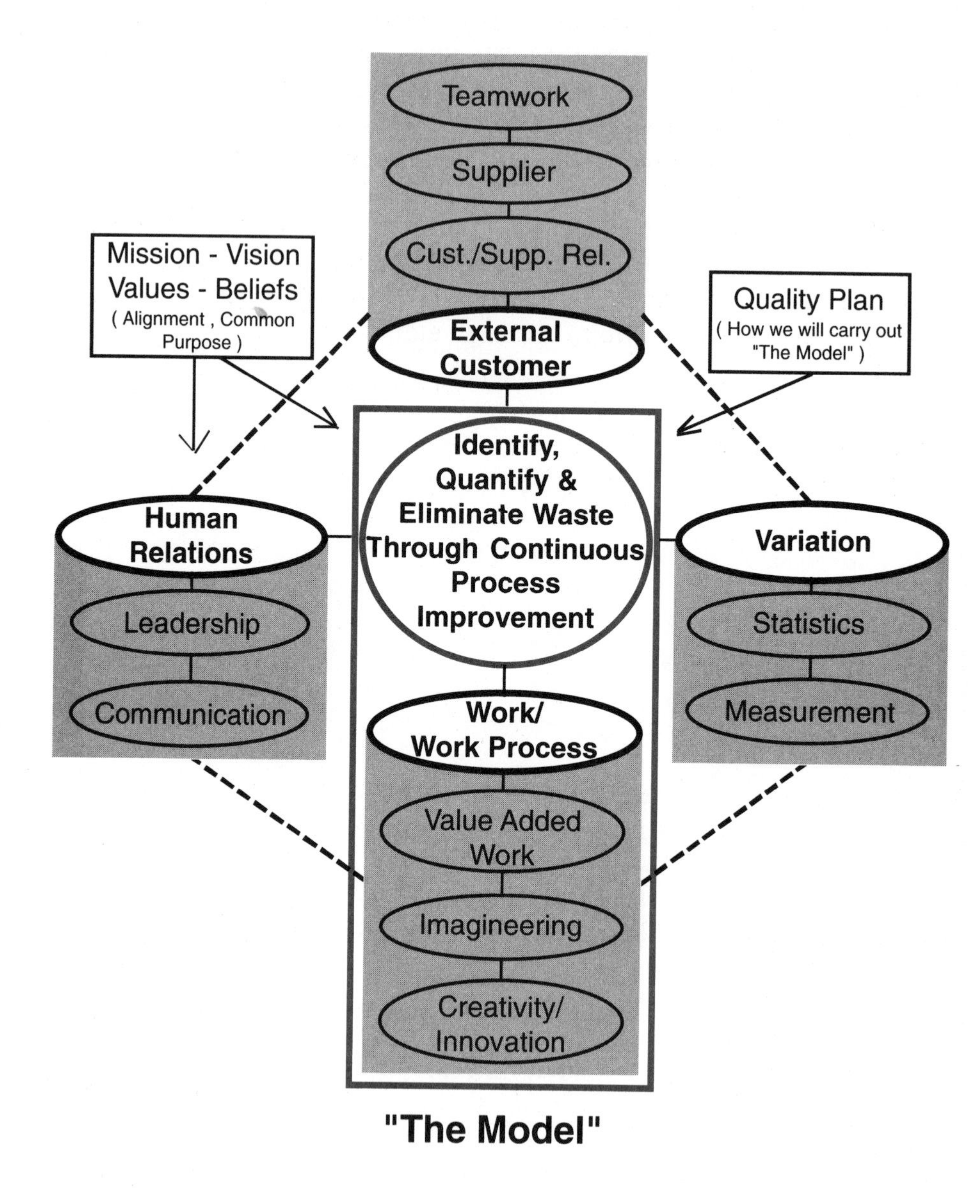
Teamwork
Supplier
Cust./Supp. Rel.
Mission - Vision
Values - Beliefs
(Alignment , Common Purpose)
Quality Plan
(How we will carry out "The Model")
External Customer
Identify, Quantify & Eliminate Waste Through Continuous Process Improvement
Human Relations
Leadership
Communication
Variation
Statistics
Measurement
Work/ Work Process
Value Added Work
Imagineering
Creativity/ Innovation

"The Model"

competitively. This cooperation extends to external suppliers and customers: customers benefit when an organization reduces its waste, and working with suppliers is an integral part of the new system. The small rectangles with arrows pointing to the central circle indicate that an organization needs a quality plan and a firm sense of its vision, mission, values, and beliefs in order to undertake the core activity. Chapters 14 and 15 explain in detail these concepts and their importance. The rest of this book provides the details, the WHATs and HOWs and WHYs and WHENs of this new system of improving quality through eliminating waste, this Right Way to Manage.

SECTION I
GROUNDWORK

If you are fully aware of how much your organization's quality could improve, how much waste it has in its work processes, and how desperately it needs to change its entire system of management, you may be impatient when you find that the first section of this book is about history and concepts and that "Getting Started" does not come until Chapter 14. In this respect this book differs from many popular books on management that claim quick fixes in months or even days. I DO think that organizations can change rapidly, but to do so, their leaders must be knowledgeable and enthusiastic about the new management system. They must understand the concepts, practice the tools, and get their eyes fixed on the goals.

Chapter 1 sketches the historical background of what I am calling The Right Way to Manage©[1], and it demonstrates with examples why I am confident this is indeed The Right Way to Manage. Chapter 2 defines three key terms. If you are unfamiliar with the roots of the quality movement, do not know the name W. Edwards Deming, and have never worked with the handful of American quality consultants who teach quality systems based on Deming's work, you should read this section with special care. It does not make sense to try to learn about, much less implement, any management system without first being convinced that the system works. If you are ready to make a commitment to quality, Chapter 1 is the place to start. It should convince you that it can be done.

Chapter 1

THE NEW SYSTEM WORKS

Does your organization judge its product quality by the day's rejection rates? Does it foster competition between its groups and individuals? Does it choose the supplier that makes the lowest bid? Does it focus only on the quality of its products? Is it confident that its waste amounts to no more than 5 to 10 percent of sales?

If you answered yes to these questions, your organization probably will not be around long. For while all of these practices and attitudes have long been accepted in 20th-Century American corporations, they will not be part of corporations in the 21st Century.

Only high-quality organizations doing high-quality work can survive global competition, but Western management has not yet realized that the science of eliminating waste is inextricably linked to the creation of high quality. People tend to attack waste haphazardly, when they trip over it. And because they trip over relatively little waste, they imagine that there is very little in their organizations. But we at Conway Quality have found that 30 to 40 percent of net sales in most organizations is wasted. Most of the waste is hidden in work processes where no one will find it unless the organization methodically searches for it, using the kinds of methods that this book discusses.

This book is different from many others about quality because our obsession with eliminating waste leads us to focus not on quality *products*—which any company can make if it is willing to discard or rework a lot of inferior products—but on quality *processes*. It will show you how to find the waste in your organization, determine which processes need the most immediate attention, and improve those processes to eliminate waste for good, lower cost, AND deliver high quality products and services. Rather than prescribe a method for infusing quality into one particular area of business, this book offers an entire system—connecting external and internal customers and suppliers, the board room, the office, the R&D lab, and the factory—as well as the tools to analyze and improve that system.

QUALITY

The two halves of this book's title—*The Quality Secret* and *The Right Way to Manage*—summarize both our goal and our method for reaching that goal. The purpose of the new system of management we advocate is to satisfy and please customers by creating low cost goods and services of consistently high *quality*. And the secret to producing high quality products and services is a management approach that emphasizes the continuous improvement of work processes through the elimination of waste.

Quality is no longer just an overused marketing buzzword or a goal to mention once a year at board meetings. Quality is SURVIVAL. Your organization will not last long if it does not use high quality processes to produce high quality products and services. Organizations around the world—your competitors—have already joined the quality revolution. But some organizations have been slow to see the light. The next ten to twenty years will reveal whether their delay will prove fatal.

The popular press has been swamped with buzzwords about quality: quality assurance, quality control, quality management, quality function deployment. This profusion indicates a lot of confusion as to just what QUALITY means. Most of us instinctively recognize a high quality product or service. It meets our functional and aesthetic needs in a superior manner. But we have more difficulty trying to apply the word to an organization. How does an organization produce high quality goods and services? We have been particularly confused over the past two decades because we have seen that the Japanese appear to be better at producing high quality than we are. Yet many Western companies that adopt "quality programs" are so slow getting results.

The apparent failure of many "quality programs" at American companies has led to a lot of criticism of American workers and management. Critics say that Americans just can not do things as well as the Japanese. But these programs do not get results because they are only *programs*. A program is not enough. To create a high quality organization requires a cultural change, a whole new system of management, a major shift in the way we think, talk, work, and act. This book provides the concepts, tools, and methods of the new system so that readers can use them to bring about such a shift in their own organizations.

Our central belief is that an *organization consistently achieves high quality and pleases its customers through continuously improving all its work processes.* Such improvement requires a systematic method of identifying, quantifying, and eliminating all forms of waste.

THE NEW SYSTEM OF MANAGEMENT

The system of management advocated by this book is not really new; it rests on sixty years of work by a number of people, not on one person's flash of inspiration. It started with Walter A. Shewhart's 1931 classic, *Economic Control of Quality of Manufactured Product.*[2] It was further developed by Dr. Joseph M. Juran, who used statistics to improve processes continuously. It was expanded into a whole philosophy of management by Dr. W. Edwards Deming, who taught it to the Japanese, starting in 1950. It was not tried throughout a large American company until 1979. Since then, it has worked in American, Canadian, and European companies, and in the U.S. operations of several Japanese companies. It is not alien to our culture, because in fact it was invented here. It can reduce waste dramatically and double and triple productivity. With the understanding and leadership of management, it can work in any type of organization.

Toyota provides a shining example of the success of this system of management. Toyota's philosophy has long been *kaizen*, Japanese for "continuous improvement." Toyota has been creating success stories for years, but one of its most dramatic triumphs came when it formed a joint venture with General Motors in the mid-1980s. As if to prove that its system could work for American companies on American soil, Toyota chose to use a plant in Fremont, California, that had been closed by General Motors a year earlier. It had been GM's least efficient, lowest-quality plant and had suffered from severe labor relations problems. Toyota agreed to hire employees from the pool of people who had been laid off when the plant was closed. Soon half the number of workers who used to run the plant were producing the same number of cars. The plant is now turning out cars of unsurpassed quality at a cost to the companies $800 per car lower than for comparable models produced by the three largest U.S. manufacturers.[3]

The new system has not been universally successful and is still news to many U.S. organizations because most companies treat it as another improvement program to be added to their present system of management. They do not recognize that it requires a complete change in the system of management, including a different mindset about how to work with employees, customers, and suppliers.

The purpose of this book is to describe the new management system, to make it understandable and practical for any organization to apply. Like most American CEOs, I too was once skeptical that any management system could produce the kinds of results that I have mentioned above. But in 1979, after a few months of working with Dr. Deming at Nashua Corporation, I was convinced. We learned to apply his principles to a variety of manufacturing, research and development, distribution, marketing, sales, and administrative processes. After this effort received wide publicity in the NBC white paper "If Japan Can, Why Can't We?," we began to work with a number of other Western companies, to help them understand this new system of management.

Because the new system requires a radical change in thinking, the usual cycle of education-motivation-action often does not work, or if it does, it works very slowly. With this system the cycle needs to be education-action-motivation. People become truly convinced of the new system's ability to produce miracles and become motivated to work in the new way only by experiencing the system's power. Action and success motivate people to further action.

Organizations that treat the new system as a program may not see overall results for five or ten years, if ever. Certainly it can take that long for a single program to change a company's culture as drastically as is necessary. But if a company aggressively takes action, creates the system, and begins to see the results, the miracles can start in twelve to eighteen months.

To take action, you need education, a system, and a plan. Chapters 2 through 5 provide a detailed description of the concepts, tools, and approaches of the new system of management. But before we get into the *what* and the *how,* we need to convince you about the *why.* And that means backing up and providing more detail about how this revolution in management got started and on what principles it is based.

THE HISTORY OF THE QUALITY REVOLUTION

Forty years ago The Union of Japanese Scientists and Engineers (JUSE) was a small but influential group helping Japanese industry. Members of JUSE were acquainted with the statistical work of Walter A. Shewhart, and some of them had met an eminent U.S. statistician, Dr. W. Edwards Deming, when he had traveled to Japan to help with the census. Others had noticed that the quality of some American World War II materiel was high, and they learned that Deming could take some of the credit, for he had taught 35,000 American engineers and technicians how to use statistics to improve the quality of materiel.

So in the Spring of 1950, Kenichi Koyanagi, Managing Director of JUSE, wrote Dr. Deming, asking him to lecture Japanese managers, scientists, and engineers on the use of statistics in quality control. Dr. Deming replied that he would be delighted to come and that he did not want to be paid for his work.

Overflow crowds of 500 people at a time eagerly attended Dr. Deming's lectures. The president of JUSE, Ichiro Ishikawa, helped Dr. Deming to arrange a meeting with the chief executives of Japan's largest and most powerful companies—Toyota, Mitsubishi, Nissan, Hitachi, Fujitsu, and 16 others. Remember that this was after five years of American occupation. Although much progress had been made, the Japanese were basically destitute—still trying to recover from the tremendous physical and psychological damage they had sustained. At the time "Made in Japan" still meant cheap, shoddy goods.

Dr. Deming's audiences knew that this man, whose message was virtually ignored in his own country, was going to talk about *quality*. But while they expected to hear about quality products, what he talked about was quality *management*. He spoke about the customer and making both the customer and supplier part of the production process. Without customers there would be no orders, no jobs, no way for the Japanese to succeed. Without close working relationships with the suppliers, no company could be sure that it would make only high quality products.

He talked about variation. After all, Dr. Deming was and is a world class statistician. He talked about how statistics can be used to identify the variation in work processes and the waste that comes from that variation. He spoke of gaining stability in a process by reducing "special causes" of variation and improving the process by reducing the "common causes." He told them how the tools of statistics could be used to improve the level of a process and narrow the range of variation.

He spoke of principles of management and human relations. He talked about listening to the workers and getting them involved in the process of continuous improvement. He explained how to treat the workers so they would perceive it was in their interest to work in this new way.

The Japanese managers did not know how to accomplish what Dr. Deming outlined, but he convinced them that they could succeed by establishing a radical new system. They heard his message about statistics, but over and above that, they heard another message. They had just lost the military war, the war of the bullets and the guns; but here was a man telling them how to win the new war, the war of industry, commerce, and finance—the war of economic power.

So they went back to their companies, not sure exactly what to do but convinced that this man had the answer to their needs. They started slowly, with one plant, one marketplace, one product, and proceeded to try Dr. Deming's principles. They gradually instilled in all their people the *will* to work this new way. Initial successes led to the *belief* that they could perform miracles. Dr. Deming's tools gave them some of the *wherewithal*, and so they *did* it. Their research identified which markets they could break into with their resources and what customers wanted in those markets. At first they chose the plums, the easy pickings: they targeted big industries with large amounts of waste. They used statistical tools to measure and control variation and to locate waste that they could eliminate.

The companies that adopted the new way of working began to see *huge* improvements in quality and productivity, *huge* reductions in cost, and *huge* increases in customer satisfaction. After five, six, seven years, they saw order-of-magnitude changes in quality and reliability—improvements of ten times, a hundred times, and even a thousand times! Employee productivity doubled and tripled.

THE CORE PRINCIPLE OF THE NEW SYSTEM

Success bred success. The JUSE continued to take the lead in training people in Dr. Deming's principles. The word spread. Japanese managers began to understand the concept of *value-added work,* work that adds value to a product or service from the customer's perspective. In five or six more years they came to a very important, though simple, conclusion. *There is little other than real value-added work in anything provided you eliminate the waste.* Although they did not state it in those words, my discussions with a number of them convinced me that this was their underlying philosophy.

That simple conclusion is fundamental to the revolution in the system of work. Think about two of the key words in that conclusion—WORK and WASTE.

When we speak of WASTE, we do not mean just the scrap and rework in the plant. We mean waste in every form: wasted material, wasted capital, wasted opportunities, wasted time and talent. Organizations waste human talent when they do not use the brains, time, and energy of all the people involved in a process. They waste opportunities when they lose a sale and the gross margin associated with that sale or when they get turned down for a grant.

Where does all this waste originate? It comes from the WORK—what people work on and how they work. Work here doesn't mean just the work of people. It means the work of machines, computers, electricity, chemicals—any kind of work.

What do we mean by *real value-added work?* We mean work that really adds value from the point of view of the *external customer* for the product or service. Most customers don't realize they are paying for a lot of waste. In fact, organizations are used to paying for the waste of their suppliers and their suppliers' suppliers, and they would be shocked at the cost drop and quality rise if their suppliers only did value-added work.

The Japanese began searching for the waste in all work—not just in the plant. They looked on each type of work as a process and examined that process for waste—everything from the design engineering process, to the sales process, to the collection of accounts receivable.

SUCCESS IN JAPAN

After the Japanese had been working in the new system for about 10 to 15 years, results began appearing in marketplaces throughout the world. People began to see different names on signs and on products, names like Toyota, Mitsubishi, Sony, Canon. German camera manufacturers couldn't understand what was happening to them. Nor could Eastman Kodak and Bell and Howell. Then came the electronic calculators. Soon companies all over the world were taking a silicon chip and packaging it as a calculator. We were all amazed when these new small calculators sold for $89.95 rather than the $600 we had to pay for the old electromechanical ones. Bowman Calculator, once the darling of Wall Street, knew it was in trouble. It couldn't survive the real world competition from companies like Casio and Sharp working in the new way. Soon calculators with higher quality and reliability and more features were selling for $12.95. The Bowmans in the business disappeared.

The steel industry provides more examples. In 1975 Congress gave the U.S. steel industry protection, only they called it a marketing agreement to limit the amount of steel imports, to give U.S. steel producers a little breathing room. After 13 years of breathing room, George Bush and the Congress felt they still needed protection and extended the agreement for three more years. Of course the industry has made progress in the conventional way of managing—cutting costs, closing plants, laying off people, and even occasionally using statistical tools. But they are far from truly accepting the new way of managing. They have to change the way they think, work, talk, and act. Their mindset must be different.

The same thing is happening in a wide variety of industries, including automobiles, VCRs, and camcorders. And the Japanese have gained advantage not only in highly repetitive operations. Consider shipbuilding, which is basically a job shop operation.

Like other shipbuilders, the Japanese usually build only one or two ships of the same design. They design the ship. They weld steel together on a simple way and move pieces with a crane to the central way. They use much of the same technology as other modern shipbuilders. And yet in five years, from 1960 to 1965, the Japanese share of the shipbuilding market went from less than 10% to more than 75%!

What is different about Japanese shipbuilding? When all the pieces of steel are brought together, they fit! The pipes, the holes, the conduits, the bulkheads, all line up. Each piece of equipment arrives on schedule and fits on its baseplate. When the equipment is started, it works. Most of the work is real value-added work, with the waste eliminated.

These changes took place 25 years ago. Yet almost no one in the rest of the world understood what was going on. Americans attributed it all to cheap labor. It was much easier to blame it on a competitors's cheap labor than on an obsolescent, complacent management system.

BRINGING DR. DEMING'S SYSTEM BACK HOME

In the 1970's, we could no longer afford to be complacent, but still very few Americans understood the revolution that was taking place. I was lucky; I was one of the few who even heard about the revolution. In 1975, I got my first inkling that something unusual was going on, but I was too provincial to take much notice. At that time we at Nashua Corporation had been working for about a year with a consortium of companies from the U.S., Germany, Australia, and Japan developing new technology for office copying machines and supplies. The copier incorporating this technology was to be built by the Japanese member of the consortium, Ricoh, Ltd. When Austin Davis, Nashua's Vice President of R&D, returned from a meeting about the copier in Japan, he was shocked. He reported that they were all crazy. They were all working on something called the Deming Prize.

"Who's working on it?" I asked him, anxious to hear about the copy machine.

"Everybody," he replied. "The engineers, the managers, the sales people, the accountants, the foremen, the production workers, the company president. They're all putting in these ungodly hours, and working like their life depended on it."

That was the first I'd heard of the Deming Prize and of Dr. W. Edwards Deming, in whose honor the prize is named. Established in 1951 after Dr. Deming gave his famous talks, it is one of the most prestigious and sought-after awards in Japan.

Ricoh did win the Prize that year. I learned more about it over the next couple of years, but I still thought it had nothing to do with me. I knew that Ricoh made one of the best engineered, lowest cost, most reliable, highest quality copy machines in the world, but I didn't connect that with the Deming Prize.

In 1979, four years after I first heard of the Deming Prize, Deming's name came up in a staff meeting. We were brainstorming about how to face the Japanese challenges to about 70% of our products when one of our vice presidents mentioned that Dr. Deming was still teaching people about quality. With Deming's help, might Nashua work as efficiently as Ricoh and produce goods as well-made as Ricoh's? Excited by the possibility, we tracked Dr. Deming down and called him that afternoon to see if he would help us. He sounded skeptical but agreed to come if I would meet with him personally.

On March 9, 1979, I and the rest of Nashua's top management met with Dr. Deming in my office for most of the day. During a brief period when Dr. Deming was out of the room, I asked the others what they thought. The consensus was that it was a waste of time. But I thought he could help us. Perhaps because of my background in numbers and work improvement, perhaps simply because I tend to perceive things in a different way from other people, I saw something in Dr. Deming that my associates didn't see and that scores of other American executives had been missing for thirty years. So despite the reservations of the others at the meeting, I asked him to work with us. Again, Dr. Deming was skeptical. He had worked with so many American companies only to see his ideas pigeonholed as a new "program" and his efforts wither for lack of top management understanding and leadership. He said he would work with us only if I personally provided active participation and leadership, made sure that the six vice presidents did the same thing, and in fact actively involved leaders at every level throughout the organization. I agreed.

Dr. Deming started working with us for two days every two weeks. *All* of our employees in New Hampshire heard his lectures. Gradually we began to understand the principles which I will expand upon in the rest of this book. Gradually we learned how to apply them in the manufacturing environment. Later we adapted them to administrative work and to knowledge work—in fact, to all forms of work.

Initial results convinced the skeptics, one by one, until virtually everyone was on board, although some of them came kicking and screaming at first. Some of the results seemed to be miracles, but more of that later.

The first few months of trying to work in this new way at Nashua Corporation generated a lot of frustration. Many people were skeptical. Others believed this was a temporary fad and gave it lip service. Even the ones who were honestly trying to make it work had a lot of difficulty understanding how to go about it. They knew I wanted statistics, so charts of all kinds sprang up all over the company. Very few knew what to do with the information they were gathering. Soon it became obvious that most of these were "charts for Bill" and were having no positive effect on the operations.

By this time, I was so convinced of the value of this new management process that I was spending 30–40% of my time promoting it. I was working with people on projects, coaching, selling, doing everything I could to persuade people that this was a permanent change. I realized it was going to take some significant successes before most of the people would be convinced of the power of working this new way.

SUCCESS AT NASHUA CORPORATION

The first major success came on a project to control the weight of coating in the manufacture of carbonless copy paper. Nashua made a lot of this paper, keeping a 96-inch wide coater running at 1,100 to 1,500 ft. per minute. So any improvement in the process was significant to the company. A too-thin coating resulted in a poor quality paper that yielded only light copies. A too-thick coating wasted expensive chemicals. The system of control was test and adjust, test and adjust. But with Dr. Deming's help, we used statistics to control the process and measure the effects of process developments, and we gradually standardized the process to eliminate variation. In only a few months, we reduced the range of weight coat variation by a factor of four! The product quality was much more uniform, and customer complaints dropped dramatically. Furthermore, a reduction in the average weight coat resulted in savings of $800,000 per year.

This dramatic success made it easier to convince others to persevere in working this new way. Even so, many took the attitude that this was a special case and the results would not apply to their operation. Eventually, however, almost all of Nashua's employees realized that they had to stop working the old way. Top management leadership and peer pressure became too great to withstand.

Another, even larger and broader success story soon began in the division that made hard memory discs for computer disc drives. This was a precision coated product that required electrical testing to find defects, since most were invisible to the human eye. We had had a large team of engineers, chemists and physicists working on improving the process for several years. Yield at final testing varied between 60% and 70% and had shown no sign of improving in spite of the many person-years of technical effort devoted to it. We took solace in the fact that our yield was somewhat better than the industry average and told ourselves that hard discs were just a tough product to make.

Once the new system of management took hold in this division, improvement projects proliferated. Hundreds of large projects and thousands of small ones studied the variation in every facet of the division's work. Improvement in the yield began gradually and then accelerated, until at the end of nine months the yield had improved from 65% to 94%. We had the same equipment and were using the same materials. The very same engineers, scientists, and production people that had worked for three years without improving the process had helped to cut the waste at final inspection from 35% to 6% in nine months. Same managers, same foremen, same production workers, everything the same. The only changes were the way people studied variation in the work and how they used that information to standardize processes and reduce waste, working together as a team. Sales, R&D, planning, everyone helped.

As we eliminated waste in the work, productivity improved—dramatically. We had set a goal of improving productivity by at least 3% per month. This meant doubling our productivity per person within two years. (1.03^{24} approximately equals two—hence a doubling in 24 months.)

When we said "per person," we did not mean just production workers, we meant every person in the division—foremen, accountants, planners, sales people, engineers, managers—everyone. We measured our productivity by dividing the number of employees in the division into the number of products (weighted for complexity) we made and sold. We wanted to double that number. The market for the product and our market share were both growing, so few people were concerned about employment security.

At first everyone thought the goal was crazy. But then as we began to see some results of the new way of managing, division management began to think that maybe we could do it in three or four years. Then, as the successful projects snowballed and added momentum to each other, the two year goal began to look more realistic. We finally doubled the productivity of the whole division in just 21 months!

Even though the market was booming, it wasn't growing fast enough to account for all of our sales increase. So we knew we had to be taking market share from the competition. We soon found out that we had the world's best marketing program, which was *higher, more consistent quality and lower costs.* The domestic competition for coated hard memory discs began to fade. Over the next several years, major companies went out of the business, even as the market continued to grow. Companies with excellent reputations—3M, Xerox, Memorex and Control Data—either sold or closed down their disc operations. They could not compete with Nashua's "marketing strategy" of high, consistent quality and low cost. Soon the Japanese too began to recognize Nashua's quality, and some of Japan's major disc drive manufacturers began buying their coated discs from Nashua, New Hampshire.

Another division at Nashua, one which had been losing money, began to work in the new way. This was not solely a manufacturing division; it provided photofinishing services through the mail to amateur photographers. Eighteen months of the new system led to quality and service improvement in all areas, while overall costs per unit fell more than 25%. Again, the "marketing strategy" of high, consistent quality and low cost began to work. Overall sales volume tripled over the next five years, and profits soared.

These dramatic successes did not come easily. Long hours were the way of life for everyone involved. There were many false starts and much wasted effort. We didn't yet fully understand just how to apply Dr. Deming's principles in a corporate environment where businesses were managed in the conventional way.

Gradually we learned how to train people, how to get them involved, how to take advantage of all the brainpower represented by our employees. We learned new ways of working with suppliers and with customers. We learned how to define projects and how to develop project teams. We formulated a system for major management-directed programs. We developed concepts of work and waste that focused on the crucial activities that improved quality, pleased customers, and drove down costs. And we helped everyone understand and communicate these activities.

Through training and experience, through trial and error, through failures and successes, a systematic process evolved. We found this process could be applied to all areas, all activities, all forms of work. It applied not only to production work, but also to sales work, research and development work, accounting work, legal work, management work, the work of machines, of electricity, of chemical processes—*all* forms of work. The *key* to improving the quality of the work processes is to *find the waste, get rid of it and prevent its return.* Improving quality through ferreting out and eliminating waste became our core effort.

Over the last decade, this process has evolved into the system that is the subject of this book. As a framework to guide your reading I will summarize in the following paragraphs the system as it is today.

THE RIGHT WAY TO MANAGE

The Right Way to Manage© is a management system for making continuous improvement through quality in all activities. The system starts with an organization's external customers who tell the organization what products and services they want, when and where they want them, and what prices they are willing to pay; it includes the organization's own operations; and it extends to the organization's external suppliers. It is the way to run a giant corporation, the U.S. Navy, a business unit, a department, a group, a team, or an individual job. The system applies to all work everywhere—the work of machines, chemical processes, computers, energy, people.

Through process improvement, the system continuously finds, quantifies, and eliminates the four forms of waste: waste of material; of capital; of people's time, energy, and talent; and of opportunities, sales, and profits. Because all waste comes from work and work processes, waste can be eliminated only by changing the work and improving the work process. Variation is the key technical tool that enables organizations to find the waste, track it down, and get rid of it through process improvement. The organization must adapt its human relations system so that people will want to work this new way. The Right Way to Manage releases the power of the people, eliminates the waste of human talent, and expands individuals' capabilities.

The Right Way to Manage provides consistent high-quality products and services at low cost to customers when and where they want them at prices they are willing to pay.

AMERICA STARTS LISTENING

As my own system developed and successes multiplied, I began to see why Dr. Deming had accomplished so much in Japan. But I was at a loss to explain why he was "a prophet without honor in his own country." I could sense that he also felt frustrated—he knew he could help American companies become world class competitors, but nobody was listening.

This American deafness began to change when NBC aired a white paper entitled "If Japan Can, Why Can't We?" This 90-minute program was first broadcast in June 1980. It included thirty minutes about Dr. Deming and his work at Nashua Corporation and an eight minute interview with me. The next day, a sea change began in American management. Dr. Deming's phone was ringing off the hook. Mine rang as well. The people calling were top management of some of America's largest companies. Companies such as Ford and Dow Chemical. Many of them wanted to visit Nashua to be convinced that this approach worked in America, in the real world. I was happy to have them come, but I insisted that the top managers themselves come, since I knew that change would require their involvement, understanding, and leadership. Dr. Deming is fond of quoting me: "...and if you can't come, send nobody."

Some of our visitors were convinced, but many still did not understand how to go about implementing Dr. Deming's principles. Most found the process so foreign to conventional management thinking that they did not believe it could be made to work in their company. In the little time I was able to spend with most of the visitors, I attempted to explain this new system of management and instill the belief that it could be done. I could see that the ideas that began with Shewhart and evolved through Juran, Deming, and the post-war Japanese businesspeople could work miracles in U.S. companies as well.

Finally, by 1983, thirty years after these ideas began the miraculous transformation of Japanese business, my personal experience had convinced me that they could work miracles in U. S. companies as well. However, the ideas were still so foreign to Western management's training and whole way of thinking that it was difficult to get people to take them seriously. Also I had found that a company adopting the new system in isolation faced difficulties; it needs suppliers who are managing in the new way. This had been brought home to me at Nashua Corporation when we tried to manufacture copy machines in competition with the Japanese. When we discovered too late that we could not buy metal parts at competitive prices, we suffered some huge losses and had to abandon manufacturing our own copy machines.

Because of this interdependence of organizations, many, many companies will have to change for the U. S. to regain its economic leadership. But since I was convinced that the new system could work for all organizations and could mean a great deal for the American economic position, I saw a huge opportunity. I wanted to be part of helping all those companies change. I retired in April of 1983 to form Conway Quality, Inc.

Its purpose was and is to educate managers as to why it is critical to adopt the new system of management, to instill the belief that it can be done, and to provide a practical way to go about doing it.

Chapter 2

UNDERSTANDING KEY CONCEPTS

In trying to focus Dr. Deming's principles of quality management into a new way of thinking, we came up with three key concepts to direct our activities toward achieving total QUALITY throughout the organization: WORK, WASTE, and CONTINUOUS IMPROVEMENT. These three concepts define the core activity of The Right Way to Manage. *The core activity is the **continuous** effort by everyone in the organization to identify, quantify, and eliminate the **waste** in all the **work** through process improvement.* This effort results in continuous improvement in quality and continuous reduction in cost, both of which please customers. In the last chapter, we discussed how these concepts came to play such a major role in the new system of work. Here we will explain them more fully and show how they come together in the core activity.

WORK

In an organization, everything depends upon what people work on and how they work. As the Japanese discovered with Dr. Deming's help, organizations should try to work on only those things that add value for the ultimate customer—the person who buys the product or service. In the new system, everyone concentrates on *value-added work*.

Value-added work increases the worth of a product or service to a customer. Customers can be *internal* (fellow employees or other departments within the company) or *external* (outside the organization). Only if work adds value for the external customer is it considered value-added work. Examples of value-added work are adding a fender to an automobile, a hamburger to a bun, or a new product to the company's line. All work that adds value helps provide a needed product or a service for some external customer. Every person performing value-added work has a customer. If it's not done to provide something an external customer needs or wants, it doesn't add value.

Some other work (perhaps 5–15%) may be *necessary* even if it does not directly add value. Filing tax returns comes to mind. But we should strive to eliminate or minimize any work that does not add value from the customer's viewpoint. Customers like to pay only for work that adds value.

Imagine if you were required to keep very detailed records of all the work your organization does and you had to analyze those records to show how much is value-added work and how much is waste from the viewpoint of your outside customer. How would you feel about submitting those records with your invoice to the customer?

You would be in for a shock if the customer cut the invoice by 40% because the customer only wanted to pay for the value-added work. Customers are willing to pay for that waste only because your competitors also have it and because they don't know it's there. As soon as customers find a supplier that has eliminated most of that waste, you will be out of business.

Think from an external customer's point of view about everything you or others in your organization are doing. Most operations are filled with activities that really are *unnecessary*. All organizations tend to maintain the status quo, and as the needs of an organization change, many activities continue that are no longer needed—if in fact they were ever needed. We keep writing reports that no one acts on or even reads. Salespeople call on low-volume customers with whom they are comfortable rather than solicit new customers.

By far the largest portion (about three-fourths) of unnecessary work is *rework*—correcting or doing things over that weren't done right the first time. For our purposes, rework includes the work of doing it wrong the first time and the inspection necessary to find the errors.

Unnecessary work, including rework, usually accounts for about **40%** of all work. Since three-fourths of unnecessary work is rework, that means rework is **30%** of **all** work.

Finally, during about 25% of payroll hours in most organizations, people are simply *not working*. This total includes legitimate time such as vacations, holidays, and breaks. But it also includes a lot of time spent waiting or doing busy work due to scheduling problems or machine down-time. And most organizations have too many payroll hours (or machine hours) for the required work. According to Parkinson's Law, "Work expands to fill the time available." In the case of production workers, Parkinson's Law often means piling up excess inventory, which is another added cost. In the case of knowledge workers, it often means generating unnecessary projects and reports, which add to cost, since other people read the reports or are brought into the unnecessary projects.

Let's put this all together to see where we stand. (See Figure 2-1 on the next page.) A typical organization expends 40% of its work on unnecessary tasks, 5–15% on necessary but not value-added work, and 25% of the payroll hours not working. This leaves *less than 30%* of the time to spend on real *value-added* work. Think of it! Only 20–30% of the time is spent on activities that add value from the customer's viewpoint. And we still haven't attempted to measure how efficiently the work is being done in that 20 to 30% of the time. What do you think the work breakdown looks like for your company, division, department—for you? Some studies have shown that only 10% of the work done in the Western world is value-added work. Is it any wonder that so many of our industries are non-competitive?

From a positive viewpoint, think of the opportunities for improvement! And luckily, the experts on the work the organization does, the ones who can help the organization sort out the necessary work from the unnecessary and make the necessary work more efficient, are the organization's customers, suppliers, and workers, the people involved in the work processes day by day. A big step towards eliminating all this wasted work is persuading these experts that it is to their advantage to help improve the work processes.

Figure 2-1 The Way We Work

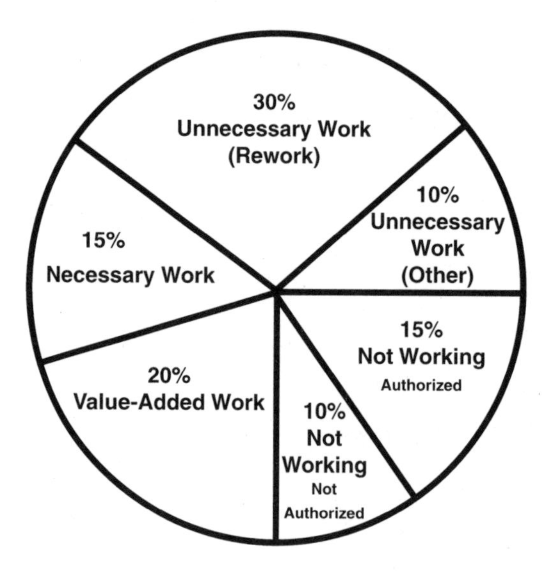

WASTE

As I learned from successful Japanese managers, there is little other than real value-added work in anything provided that you eliminate the waste. This sentence ties work and waste together. As we said earlier, waste comes from what people work on and how they work. Most work that does not add value to the customer is a form of waste. Even if people are engaged in value-added work, but are doing things wrong the first time or inefficiently, they are causing waste.

Organizations waste materials, capital, people's time and talent, and the opportunities that could lead to sales and profit. Material waste includes waste of energy as well as scrap and rework and waste caused by over- and under-specification of materials and finished products. Remember organizations need to produce what adds value from the customer's viewpoint—no less and no more.

Waste of capital includes capital tied up in excess inventory, receivables, plant, equipment, and other assets. Excess inventory is particularly damaging since it also creates additional cost for space, handling, insurance, damage, and obsolescence.

People in organizations waste time, from the CEO who spends the week going to unnecessary meetings to the stock clerk who must kill time waiting for the next delivery. And every organization wastes people's talent either because the wrong people are in the wrong jobs or because of all the unnecessary restrictions that organizations place on employees' activities and creativity.

To prevent waste from lost sales, organizations must present the right product(s) to the right customer in the right way and at the right time and price. Including the internal customer in the previous sentence makes clear that this kind of waste can occur in every department, from research, through product development, marketing, and finance. Sales people aren't the only ones who miss opportunities and lose customers. Some of the greatest waste comes from not identifying customers' needs and not having the right product to fill those needs at the right time. The customer is key in avoiding waste since the customer's wants and needs determine value.

CONTINUOUS IMPROVEMENT

As we saw in the first chapter, some companies have adopted and then abandoned quality programs because they treated them as just *programs.* A company cannot eliminate all the waste from its work processes once and then assume that from then on, all its work will be value-added work. When we say eliminate the waste, *forever,* we mean this is a *continuous* process, a permanent part of a corporation's culture. All people at all levels have to change the way they think, talk, work, and act. Continuous improvement in quality and productivity must be "job one" for everyone. Continuous improvement is a relentless, never-ending process.

This change is so difficult because it's never over. Rather than a one-time project to make improvements, it is a whole new way of operation which requires a major culture change. It is a theory-shattering shift in the way of thinking about an organization. It requires the complete understanding and the active leadership of top management. Otherwise it will not last. Maintenance of the status quo is a powerful motivator for most people. To make a permanent shift in the way people think, talk, work, and act requires dedicated leadership over an extended period. This journey should not be embarked on lightly, but the rewards are so great and the alternatives so painful that we really can't afford not to begin.

SECTION II

USING THE TOOLS AND APPLYING THE CONCEPTS

Organizations get rid of the waste in their work processes by applying the concepts introduced in the last section and using the tools of The Right Way to Manage, including work analysis, variation, charting, and imagineering. Imagineering comes last in the sequence, even though it is the most important tool, because work analysis and charting provide the facts upon which to base imagineering.

This section explains how to apply the concepts discussed in Chapter 2, introduces the tools, and then puts the tools to work in the war against waste. Chapter 9, the last in this section, presents a process improvement methodology, an approach that draws upon everything else in the section.

Chapter 3

ANALYZING WORK

In its February 23, 1980, issue, the English magazine *The Economist* stated, "The main secret behind any economic or social advance is the simple one of a relentless daily productivity hunt. The dividing line between successful and disastrous organizations in the world today is between those where the working force plans hourly for greater output tomorrow (call these enterprising) and those where most people are concerned to avoid any bothersome disturbance tomorrow (call these bureaucratic)."[4] This quote describes a major part of the core activity of The Right Way to Manage, the relentless daily pursuit of continuous improvement. One of the most fruitful tools in this daily pursuit is the analysis of work.

In Chapter 2 we discussed the concept of work and why it is so important, and we said that people spend their work day engaged in six kinds of activities. Figure 2-1 on page 24 indicates the percentage of working time devoted to each of the six categories in most organizations.

Obviously, an organization whose people spend less than 30% of their time on value-added work needs to find and eliminate waste in its work. As a start, everyone in the organization needs to think of work in these six categories. So much work is wasteful in part because people usually equate being busy with doing work that has value. Very few people look for the waste in knowledge work and other work processes that don't contribute directly to production.

When management is encouraging people to look at their work with fresh eyes, resistance to change is one of the hardest attitudes to overcome. But gradually people need to replace it with an attitude of constantly looking for change. The organization should reward people who look for change and are creative in eliminating waste. Too many corporate bureaucracies reward those who supervise the greatest number of people. Is it any wonder that such a system perpetuates tasks that add no value? It is human nature for people to want to believe that the work they are doing has value. Nothing is more damaging to self-esteem than to work continually at tasks that have no value to anyone. Insuring that everyone is doing necessary and valuable work improves morale.

To function in the new way, organizations should study work in order to maximize time spent on work that adds value and minimize or eliminate work that does not add value. They should determine what work they really pay people to do and how long it takes them to do it.

The first step in applying these basic concepts is to analyze in great detail all the work presently being done. Organizations must understand what tasks are being performed before they can eliminate the waste. The object of the job analysis is for management, as well as the people doing the work, to understand what is being done, how it is being done, and the cost and value of those tasks to the company.

An analysis of work determines the value of the work and work process from an overall company point of view. It breaks down that work into bite-sized pieces and examines work processes to see what can be eliminated, combined, reduced, or done in an better way. The advantages of this approach are many. While reducing waste, it improves quality. It is relatively easy to gain consensus for this approach because people know the waste is there. They understand the organization is taking a logical approach to eliminate it. They don't want to lose their jobs, but they do want a full day of meaningful work in a company that is competitive.

As we define it, work is a set of tasks performed by people, machines, energy, chemicals, etc., to meet an objective, and it is measured in terms of the time it takes, what it costs, and the quality of the work product. To measure the time taken, the organization breaks down the work into tasks and sub-tasks. To measure quality, the organization picks a method appropriate for the task being measured. Calculating percentage of errors is a good method for many clerical operations. It takes ingenuity to find the best quality measure for types of work that do not have error rates, rejection rates, or the like.

To analyze work, divide it into the six categories shown in the pie chart. Obviously, the objective is to eliminate the unnecessary work and unauthorized non-working time, minimize the time spent on purely necessary work, and improve the quality of value-added work by reducing waste during the time spent on it.

Since much "necessary" work that does not add value gets performed for internal customers—people or groups within the same organization—analyze closely whether the internal customer really needs something from the point of view of the whole company and the external customer. Consider what things the company really pays employees to do.

Figure 3-1 lists examples of the six categories of work:

Figure 3-1 Categories of Work

***VALUE-ADDED WORK**

Working on the right things
Doing the right things right
Doing the right things right most of the time
Working at the right pace

NECESSARY WORK

Filling out expense reports
Approving expense reports
Traveling between plants
Filing tax returns
Taking physical inventory

UNNECESSARY WORK (Rework)

Making errors
Finding errors
Fixing errors
Complexities
Interruptions
Inspection
Customer complaints
Redesign, etc.

UNNECESSARY WORK (All Other)

Preparation of reports that no one uses
Tasks that are part of system but no one wants/needs
Reading reports or attending meetings of no value to you
Working on the wrong things

NOT WORKING - Authorized

Breaks
Holidays
Vacations
Personal time

NOT WORKING - Not Authorized

Waiting time
Idle time
Breaks
Arrive late
Leave early

* This is the only kind of work our customers pay us to do!

The definitions of the kinds of work—value-added, necessary but not value-added, and unnecessary—are also extremely important to people's thinking about work every day. Rather than think about work as the set of tasks being performed, people should continually examine those tasks so that they automatically classify them into categories.

Approach the analysis by asking a series of questions:

Is the task necessary for the company?
What is the task's value for the company?
Is the work being done by the right part of the organization?
Is it being done by the right people?
Are the best possible processes being used?
Can we improve the processes?

Question, question, question, always keeping in mind the objective of eliminating all forms of waste—the waste from unnecessary tasks and the waste from important work not being performed as it should be.

This approach can be very eye-opening in exposing the size and the types of waste. One company had a cost department that computed job order costs for one of its divisions because years before the vice president in charge of that division decided the manufacturing people needed the actual cost on each job to monitor and improve their performance. The controller responsible for the cost department felt the information was useful for valuing finished goods and assumed the manufacturing people found it valuable.

Unfortunately, the costs couldn't be accumulated until, on average, thirty days after the job had been shipped. Very little could be learned at that point, and the manufacturing people had developed other real-time measurements to control their operations. For example, they used pounds-finished-per-labor-hour as the primary measure of labor efficiency. The manufacturing people largely ignored the voluminous cost reports and assumed they were needed for accounting purposes. A study of value-added work showed that the expense of the cost reporting system was not justified from a total company point of view.

The company eliminated the cost reports by establishing a simple system of estimating inventory value, freeing three people from the cost department. One went to work for the manufacturing manager analyzing and investigating problems of excess cost. Another became a cost estimator, needed because the company had been underbidding many jobs as a result of sending out quotes without first making an estimate. The third one replaced someone retiring in another department. In their new jobs, these three people contributed much more to the company.

It sounds simple, but because organizations seldom examine the value of work being done and departments often communicate poorly with each other, effort is wasted all over the place. Waste accumulates over time and grows as the organization gets larger. People who go through this rigorous analysis are often amazed at the amount of waste.

The problem, of course, is management. Management doesn't know enough about work and work processes and doesn't provide the leadership required to generate change.

DIRECT AND NONDIRECT LABOR

Historically, Western industry has dedicated itself to controlling *direct labor* costs, the costs of labor that directly contributes to the production of goods or services. Frederick Taylor's time and motion studies and other industrial engineering techniques focused attention primarily on the work of direct labor. *Nondirect labor*—work that does not directly contribute to production, including most types of staff, administrative, and knowledge work—now occupies more employees than direct labor, and the waste in nondirect areas is usually many times larger than in direct labor areas. However, the work done to eliminate the waste of people's time continues to concentrate on direct labor.

In Japan, where major strides have been made in increasing productivity of direct personnel, the proportion of costs that are nondirect have been rising dramatically. For firms in Japan employing more than 1,000 people, the percentage of employees doing nondirect work has increased from 22% in 1965 to 43% in 1978 and to approximately 55% today.

In the U.S., the trends have been similar. Last year, approximately 30% of the non-agricultural workers in the U.S. did direct work. Clearly, nondirect work deserves much more scrutiny and should be the major target when organizations hunt for waste. It is responsible for more than twice the hours, and because of pay rates, probably three times the costs of direct work. The tools are available to identify and eliminate that waste and managers cannot afford to neglect it if their organizations are to survive as viable competitors.

Why has nondirect labor grown to absorb such a large proportion of employees' time? Although much remains to be done, many companies have spent substantial time and effort reducing waste in direct labor. Yet in too many companies, bureaucracy and bureaucratic practices are spreading. Webster defines a bureau as "a business establishment for exchanging information, making contacts, and coordinating activities." How many people in your organization, including yourself, spend a great deal of time exchanging information, making contacts, and coordinating activities? Yet how much of that activity adds value to the organization's products and services? Sure, a lot of it is necessary, but in most large organizations, it has grown out of all proportion to the need.

One company we're familiar with increased its production volume 35% over five years. Meanwhile, the number of direct labor personnel went up 4% but the number of nondirect and salaried personnel increased 44%. The company is no longer competitive, even though it made a substantial improvement in the productivity of its direct labor force. It paid a great deal of attention to improving the productivity of direct work but little, if any, to improving nondirect work.

Because of the constant emphasis on costs, direct workers are often very conscious of the company's need to control costs. On the other hand, most nondirect people do not understand costs or are not conscious of the pressing need to control them. Any direct worker knows that if she produces forty widgets in eight hours rather than fifty, those widgets are going to cost the company more. Furthermore, if ten of the widgets turn out to be faulty, the costs will be higher still. But the cost of their work is not so obvious to the product manager or to the clerk keeping inventory records. Their feeling often is, "What doesn't get done today will get done tomorrow." They don't think of mistakes as costing the company money, since they eventually get corrected. Because nondirect jobs are so diverse, they are difficult to analyze. Often the person doing a job is the only one who knows all the tasks the job includes, and he or she usually does not understand all the reasons for the tasks from a total company viewpoint.

People doing nondirect work seldom feel pressure to reduce costs. Historically, their employment has not fluctuated with the fluctuating work load. Also, nondirect workers usually have no inventory to dampen fluctuations in their work load. Consequently, Parkinson's Law is very strong in nondirect areas; the work expands to fill the time available. Companies hire staff to meet peak loads with some cushion thrown in, giving little consideration to the costs versus the benefits of staffing for a peak load.

In the direct areas, flexibility of the work force and of scheduling are critical in keeping costs under control. A manager of direct personnel knows that if he or she has more people than necessary for today's work, costs will go up. In nondirect areas, staffing and scheduling are just as critical to costs. However, since cost is not paramount, managers in nondirect areas often make staffing and scheduling decisions to avoid criticism of not doing the job on time. Most cost reduction effort still goes on in direct areas because it is easy to understand and measure. Peter Drucker said in a Wall Street Journal article: "If you're working on improving labor productivity, you're wasting your time. Very few companies have more than 10% labor costs."[5] He was referring, of course, to direct labor costs.

WORK SAMPLING

Before the job analysis can start, the organization must determine what work is actually being done and how people are spending their time. One tool for accomplishing this is the flow chart which we will discuss in Chapter 5. A second important tool is work sampling, a fairly simple procedure that gives a reasonably accurate indication of just what work is being done. To sample work, take a large number of observations at random times over a full cycle of the work and note what activity is being performed at observation time. Individuals can do their own sampling. Past perceptions of work sampling studies leave many with the fear that a stopwatch will be put on their every move and that an "efficiency expert" will scrutinize their performance. Work sampling is handled differently in the new system. It is largely participative and thoroughly explained in terms of its benefits to the people involved.

New electronic random timing devices allow an organization to sample work with participation from everyone. When work sampling is handled correctly, no one looks over people's shoulders, and no one objects to it. All employees can see that they are helping to collect data that will make their own work more useful and rewarding, and they are not being personally evaluated or criticized. This way it is relatively easy to gain acceptance and even enthusiasm.

The people doing the work often have a difficult time understanding what they spend their time doing versus what they are paid to do. This fact was highlighted for us during a year-long assignment we had at Portsmouth Naval Shipyard, employing 8,000 people. At one point I was interviewing some of the unskilled workers, including three sandblasters.

I asked one of the sandblasters, "What are you paid to do?"

He replied, "Shoot the gun."

The sandblasting gun scours the sludge, dirt, grit and rust from the hulls of ships being refurbished. When asked what percentage of the time they actually spent shooting the gun, the sandblasters estimated 60% or 70%, but they said they could not be sure. The workers then used work sampling to determine about how much time they actually spent sandblasting, the work the Navy paid them to do. After they analyzed their own work, they were amazed to find that they spent only 30–35% of their time on this activity. The work sampling revealed much about what activities made up the other 65–70% of the sandblasters' time and pointed the way to improvements in the work process. The workers knew what they were supposed to do. Now management had a tool to find and remove the barriers to productivity in the system that prevented the workers from doing value-added work.

Organizations have an even more pressing need to use work sampling to discover how professional employees spend their time. Knowledge workers perform many diverse activities, and most have only a vague feeling of how they apportion their time. Many of them feel that they must often neglect important parts of their job to perform a myriad of seemingly urgent yet insignificant tasks. "We often neglect the important for the merely urgent."[6] When managers, even CEOs, do work sampling of their own activities, they are almost always shocked at how much time they waste on unnecessary and/or insignificant activities. But without detailed knowledge of how they actually spend their time, they have trouble making significant improvements. Remember Figure 2-1 on page 24 categorizing activities while at work.

People at any level can perform work sampling. The best way is usually to have people sample their own work, which they can do simply and effectively with the form shown on the following page and an electronic random timing device.

The person categorizes his or her work activities across the top of the page. The timing device buzzes or pulses at random intervals, and the person makes a check mark under the activity he or she is performing at that moment. After a large number of observations, it becomes apparent how a person is spending his or her time. The organization can use the information gained from the work sampling to get an idea of how many person-hours the entire organization is devoting to various activities and ultimately to identify and eliminate waste.

QUESTIONS TO ASK ABOUT WORK

Work sampling, task analysis, time and cost estimates, and the charts that we will study in Chapter 5 provide the raw materials—facts—for imagineering how things could or should be, the subject of Chapter 6. As preparation for imagineering about work, people from all levels of the business unit brainstorming together should come up with ideas for eliminating unnecessary work and for eliminating waste in the remaining work while improving quality. If possible, someone from outside of the unit should contribute ideas and suggestions. An outside objective viewpoint can be very useful.

As a start, pose questions for each major activity and for each detailed task. Be receptive to all types of ideas. Look for major as well as minor changes in the way you operate. Imagine if you did not have any organization and were starting fresh. What is the real purpose of each activity? What tasks are required to ac-

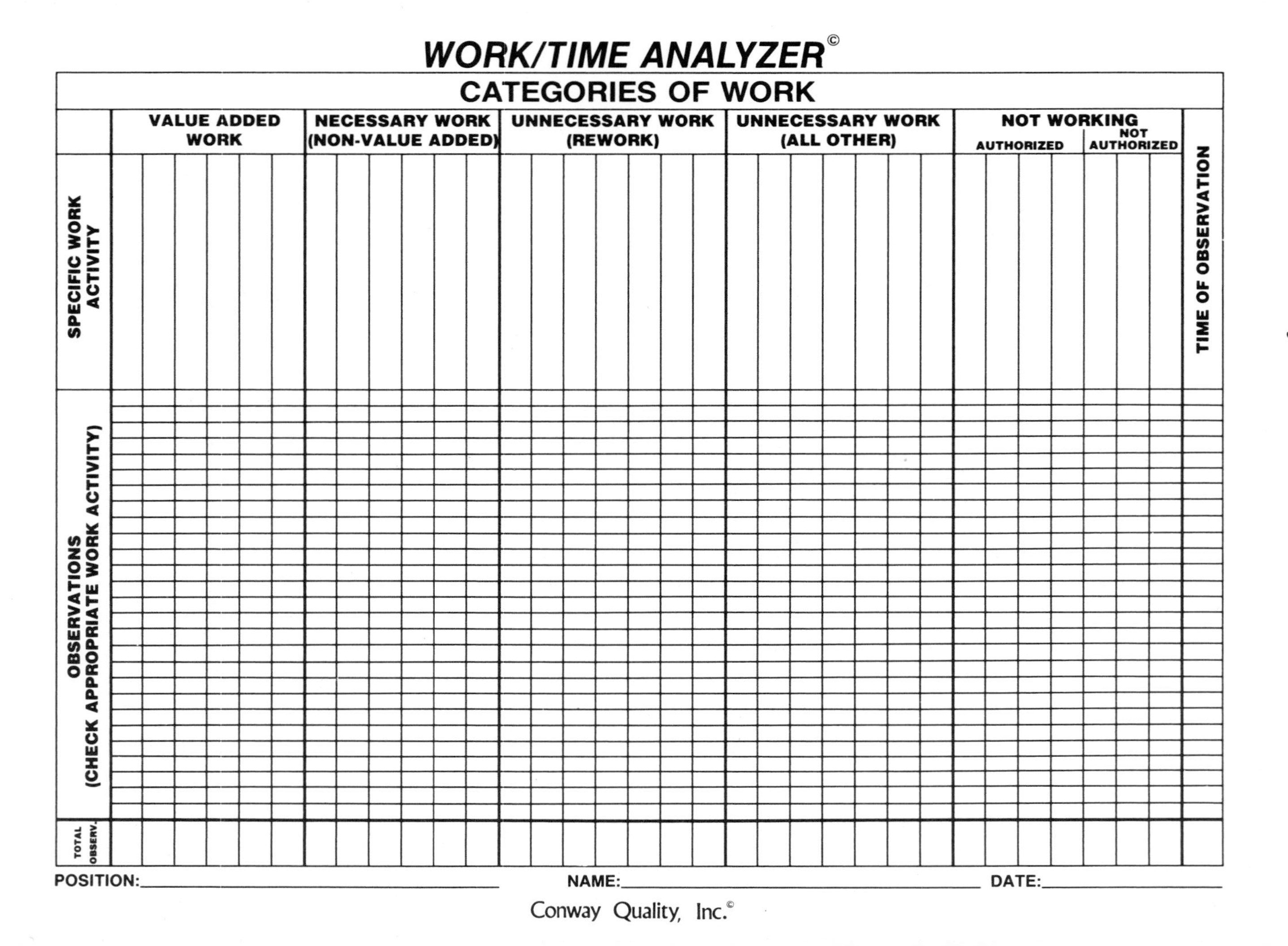

WORK/TIME ANALYZER©

CATEGORIES OF WORK							
	VALUE ADDED WORK	NECESSARY WORK (NON-VALUE ADDED)	UNNECESSARY WORK (REWORK)	UNNECESSARY WORK (ALL OTHER)	NOT WORKING AUTHORIZED	NOT WORKING NOT AUTHORIZED	TIME OF OBSERVATION
SPECIFIC WORK ACTIVITY							
OBSERVATIONS (CHECK APPROPRIATE WORK ACTIVITY)							
TOTAL OBSERV.							

POSITION:______ NAME:______ DATE:______

Conway Quality, Inc.©

complish the purpose of each activity? What organization, systems, and processes would accomplish the purposes most effectively? Assume that you only want to do those things that are worth more to a customer than they cost you to provide.

Ask a lot of questions about each activity and task, questions like these:

What is its worth to the customer?

Discuss with both internal and external customers which of the services you are providing they really need and which they can do without. I know of one company that called a particular customer every time an order was shipped to tell that customer when and how the order was being shipped. When asked, the customer said the service was unnecessary; apparently this task arose from a one-time request made two years previously. The shipping company entered the request under "special instructions" on the customer's master file, and no one ever questioned it. This type of thing can go on forever unless the organization uses a methodical system to analyze and question the value of tasks being performed.

Is the purpose of this task still valid?

This question usually uncovers a great deal of waste. Tasks are often performed because "We've always done that." People at many levels do things because they assume that the boss wants them to or that someone needs them done. Often, however, the need has long since disappeared, if there ever was a need in the first place.

What risks would eliminating it pose?

One large Japanese company carried this question all the way to the Board of Directors. The Board required a revised budget every two months. Analysis showed that these revisions cost the company $700,000 a year. The company examined the possibility that its decisions might not be as good without the budget revisions and decided that the risks were small and it could get by with one semi-annual revision, saving over $400,000.

On a more mundane level, what are the risks associated with eliminating credit approvals on all orders under $500? or $1000? or $5000? The answers would vary, of course, with the type of customer base a company has, but these kinds of automatic tasks are worth re-examining periodically.

Can you transfer it to someone else outside the company?

Many companies have their data entered and forms prepared by outsiders, to everyone's benefit. More and more suppliers and customers are talking to each other via computer, rather than through paperwork. Although suppliers and customers are the logical people to consider when imagineering the transfer of work, subcontractors are also good possibilities. Many times, specialists can do a job much more cost effectively than you can. Auditing of freight bills is a common example.

Can you consolidate it with some other job within the company?

Consider the order entry function, which is usually performed by a clerical group taking salespeople's orders, checking them, and entering the information into a computer. Many companies now provide their sales force with portable computers to enter orders. The software program has many built-in checks, and the completed order can be sent over the phone line directly to a host computer. This eliminates two steps in the process and often improves accuracy.

Can you lower the frequency of the service?

Ask this question and the next three of all reports prepared within the company. Reports seem to have a life all their own. Once the seed of an idea for a new report is established in someone's mind, it inexorably grows until the report is born. Reports are rugged survivors. They don't often die of accident, sickness, or even old age. The attempt to kill a report is often treated like an attempt at murder. The time it takes to prepare useless reports is only part of their wastefulness. Reading them wastes even more time, although such waste is difficult to pinpoint.

Obviously, the question of frequency applies to services other than reports. Consider everything from the frequency of trips to the post office to the frequency of Board of Directors meetings.

Can you reduce the content of the service?

In a division of a large manufacturing company, a monthly accounting report on capital expenditures listed every single project, whether it was the purchase of a desk or of a million dollar piece of manufacturing equipment. It also included completed projects until the end of the fiscal year. By dropping completed projects and those under $10,000 from the report, the company reduced the size of the report by three-quarters and made it much easier to use and to prepare.

Can you reduce the number of people or groups receiving this service?

Once you have printed a report, the easiest thing in the world is to make more copies. The extra paper doesn't add much cost. But the copies may waste a lot of time of the people who read it but take no action or make no decision based on the report. To avoid such waste, always ask, "Who needs to know?," not "Who would like to know?," or "Who would it be nice to inform?" The waste here is much greater than you realize.

Can you change the methods to make the job more efficient?

This is obviously a major ongoing question. For the first round of improvement after analyzing all the activities, look for obvious improvements in methods and systems. Later, use the tools of variation and the simple industrial engineering concepts to refine methods.

Can you mechanize it or computerize it?

This is a very important question, particularly with the wealth of computer power available at lower and lower cost. However, we must put computerization into proper perspective. It does not, in and of itself, get rid of wasteful work. It makes all work more efficient, including wasteful work. This question comes last for a very important reason. Too often, we computerize wasteful effort and thereby merely do unnecessary work faster.

The key words in these questions are **eliminate, reduce, transfer, change, combine, rearrange, simplify.** These questions need to be asked over and over again of each activity, task and sub-task. A number of people need to be involved in answering the questions—unit managers, the people doing the work, internal and external customers, other related units, and outside independent consultants who might contribute ideas. Openness is critical. All must understand that everyone is granted amnesty for the past and that mistakes in the future will also be given *amnesty,* as long as people are trying for continuous improvement. The new system requires change, and change involves risk. People should be able to risk making changes with uncertain outcomes without worrying that they will be criticized for their initiative if things do not work out as planned. Otherwise initiatives for continuous improvement may be stifled. In The Right Way to Manage we blame the system or process—not the individual.

WHY AREN'T YOU WORKING?

In addition to the types of work, examine closely the time spent not working and the causes of that time. Time not working can be broken into two parts:

1. Officially authorized vacations, holidays, breaks, sick leaves, wash-up time, etc. and
2. Other or unauthorized—time spent idling, waiting in line, awaiting instructions, leaving early, daydreaming, extending coffee breaks.

Managers may sell their employees short by assuming that people spend too much time not working because they are lazy or not motivated. While this possibility can't be ruled out, time spent not working more often results from company culture and scheduling and staffing problems. These are a part of the system, the process, "the way we do things around here." Gradually, over a long period of time, the company culture develops concepts of what is a fair day's work and how hard one should work. These concepts change slowly over a period of time, and if no one does anything about it, the amount of work required for "fair effort" may slowly dwindle. Management can improve the average effort by leadership and example. If the CEO gets to work on time every day, the people next in line are likely to do the same, and so on down the hierarchy.

The company culture also determines the pace at which people work. This pace tends to slow over time unless management makes known what constitutes a fair and acceptable pace. Much work has been done to establish a fair pace that can be sustained without fatigue in repetitive operations. For non-repetitive operations, leadership and example must set the pace.

In most cases, though, company culture is not the major cause of lost time. Within wide limits, the productivity of nondirect personnel is determined by two factors: 1) the amount of useful work to be done, and 2) the number of people to do it. Dividing one by the other yields a simple conceptual measure of productivity, the amount of useful work per person.

Many factors create a tendency for the number of nondirect people doing a given job to increase. Managers want their divisions to do a good job and therefore always want a little insurance, in the form of having extra workers available, to be sure the division can perform quickly and well. Many nondirect tasks tend to be cyclical, with a peak of activity at certain periods. With pressure to improve performance in the form of meeting tighter deadlines, the peaks of effort tend to increase, requiring more manpower to meet those peaks. A manager's status is usually judged, in part, by the number of people working for him or her. Both to increase status and to insure the job is done thoroughly and on time, managers tend to want to add people. Over a period of time, wasteful tasks tend to become part of the process, because there is no mechanism for a methodical examination of the necessity and cost of tasks being performed versus their value.

When time is available due to overstaffing or scheduling problems, work expands to fill that available time, whether it's useful or not. If the extra available time is not spent with make-work, it is spent waiting for work. Such time is a huge waste, from the board room to the factory floor. Usually, however, it is an accepted way of operation. In most companies, time spent in this manner varies from 10% to 30%, but it can go as high as 50% or 75%. If it is 10%, improvements require a lot of effort. If it is 20–30% or even higher, dramatic gains are usually easy to make. What do you think this figure is in your company? To your guess, it is probably safe to add 10–15% to arrive at the true figure.

Examine schedules critically to level the workload as much as possible. Try to identify value-added work that does not have critical deadlines and thus can be used to fill in the valleys after peaks caused by tight deadlines. Mail order photofinishing has a particular cyclical problem because of seasonal peaks and peaks on the couple of days following a weekend or holiday. Nashua Corporation's photofinishing division identified reprint and enlargement work as tasks that could have a more flexible schedule. It even contracted to do outside data entry work without tight deadlines. Despite these steps, the division couldn't even out its workload, so it arranged for a flexible work force, of both direct and nondirect people, consisting of a base complement of permanent workers supplemented by part-time workers to meet the peaks. The result was that people on the job had full-time meaningful work available for them. Altogether, these changes helped reduce the division's unit labor costs by 27% over a 15-month period. And because the division could reduce its prices and thus obtain substantially more volume, the actual number of employees went up.

OTHER KINDS OF WORK

The analysis of work should extend beyond the work of people to include work done by machines, computers, chemical processes, electricity, etc. The same principles apply: eliminate the waste and keep only work that adds value. For example, the work done by electricity in lighting a room is not value-added if no one is in that room. Similarly air conditioning of unoccupied areas does not add value. Gathering facts on energy use and then analyzing the work done can be an effective way of cutting costs.

Computers can do a prodigious amount of work, but the work they do often contains a great deal of waste. Organizations seldom examine this waste because of the general feeling that anything done by the computer is "free." Computers are extremely valuable tools, but only if they are made to work on the right things. The more work computers do in an organization, the more important it is for the organization to analyze the work to determine what portion of it is adding value.

A chemical cleaning process I once studied had been doing satisfactory work in that the cleaned parts were always clean enough for subsequent processing. However, by examining the process for waste and analyzing what really was necessary (what added value), we discovered that the process used more time, chemicals, and heat than were necessary. Adjusting these three elements produced a substantial reduction in cost, yet the process continued to get the parts sufficiently clean. Remember, the ideal process not only works every time, it works without waste.

The analysis of work is not a one-shot program to obtain improvements in operations. It is the first step, and it provides the basis for continuous improvement. This analysis yields a lot of knowledge about the work and work processes—what the work is and how it is conducted. To gain maximum benefit from this knowledge, the organization must operate on new assumptions. The key assumptions are:

1. Continuous improvement is a way of life.
2. The customer's requirements are key to everything the organization does.
3. The organization will eliminate rework and other unnecessary work, minimize time not working, minimize work that is necessary but doesn't add value, and continuously endeavor to improve the way it does value-added and necessary work.
4. The organization must use the tools available to analyze the work in more detail to accomplish Number 3.

When analyzing work always remember that the work not done could be the biggest waste of all. That is the work the external customer needs, wants, and is willing to pay for because it adds value. The errors of omission are often bigger than the errors of commission.

Everyone should be conscious of the cost of providing services. Cost consciousness should pervade the nondirect operations as well as the direct production activities. After working the new way for a period of time, everyone at all levels will automatically think of work in its six categories. When people continuously think of whether the work is necessary and what its value and costs are, continuous improvement will follow. People want to do worthwhile things. They need the tools and concepts to help them determine what is worthwhile. Classifying the work into its six categories is one of the most powerful ways to begin.

A key attitude for continuous improvement is that the organization is examining work and waste in the process, not people's performance. By concentrating on the work being done, the organization can make value judgments about the things being done, not about the people doing them. Since the system, not the people, causes most of the problems, concentrating on the system is much more fruitful than focusing on the people. It also elicits cooperation rather than antagonism. While almost everyone is concerned about employment security, people seldom feel threatened by efforts to improve work processes as long as they understand and participate in those efforts.

Most important of all, the reward for this effort is a new competitiveness. Everyone recognizes the need for low costs and high quality in the product or service the company produces for its external customers. Today, that is not enough to insure competitiveness. The people producing services for internal customers must also have high quality and low costs. Everyone in the organization needs to do high quality value-added work without waste—now, tomorrow, and forever.

Chapter 4
PROCESS VARIATION

This chapter introduces the concept of process variation and its uses in the drive to eliminate waste through continuous improvement. Process variation is such a central part of The Right Way to Manage that virtually every subsequent chapter will expand and refine the ideas in this one.

A PROBLEM AND A SOLUTION

Variation is "the change or deviation, in form, condition, appearance, extent, etc. from a former or usual state, or from an assumed standard." Process variation—the deviation from a norm within a process—is both a leading contributor to waste in work processes and one of the principal tools organizations can use to find waste and eliminate it. Once people know how to understand or read process variation, it can act like a map pointing the way to the parts of the process that need most attention. That is why I describe variation as "the guide, the compass that tells people which way to go, what to work on to effect continuous improvements under the new management system."

It's no secret that the variation in a work process causes quality problems. If one year a technical college gives degrees only to mechanics who can reconstruct an engine on their own, and the next year the same program graduates anyone who can pass a written test, the "customers" of that process—the businesses that hire the mechanics—may be very unhappy.

There will always be variation in every process, so we will never reach perfection. No two things are exactly alike. Provided we use accurate enough measuring instruments, we will always find differences. If differences do not show up, we are not looking closely enough—or the measuring instruments aren't sensitive enough. But we can strive to reduce the range of variation. We can also improve the average process level, the norm around which the variation takes place.

So besides being the source of waste, variation is the technical tool with which we find problems, errors, complexities, and waste. It helps us track them down, get rid of them, and prevent their return. We gather data on the variation and turn that data into useful information by using several technical tools. The whole system of continuous improvement is based on using variation.

Analyzing process variation is the first step in meeting a customer's needs. If an organization understands a customer's requirements and reduces them to operational definitions, it can use its analysis of its own work processes to determine whether it is capable of meeting the customer's needs. If the present process cannot turn out products and/or services that meet the customer's requirements, study of the process variation can help the organization change the system to improve the process.

We record the variation in the work results, the work process, and the waste. When we record the data and use statistical methods to convert the data to information, we learn what to do—what to fix, what to eliminate, what to improve and how. Using statistics requires measurement. Something that can't be measured is difficult to improve. Measurements tell us how we are doing, what our results are, and how the work processes are delivering those results. We need to understand the work and work processes thoroughly to determine what to measure and how to use the measurements.

Unfortunately, in the old system, variation is like a fog obscuring our understanding of what is happening in a work process. As Dr. Lloyd S. Nelson, Director of Statistical Methods at Nashua Corporation, says, "We need to become *masters* of variation, instead of victims of variation . . . . *The central problem in management and leadership is the failure to understand the information in variation.*" [7] Variation contains a great deal of valuable information, if we learn the tools to uncover and communicate that information.

GETTING CLOSE TO THE PROCESS

To see variation in a process, you have to get close to it, probably closer than you have ever been. Imagine that you are walking toward an orchard. From a distance of 100 yards, all the trees look the same. You really see no variation. Then as you approach, you see that one group of trees has a different shape. Still closer, and you see that some are apple trees and some are pear trees. From 20 yards you observe that the pears on the sunny sides of the trees are ripe, while the ones in the shade are still green. One tree has fewer pears than the others, the pears are much smaller, and the tree itself looks sickly.

You pick two ripe pears from two different trees and take them to the laboratory. You measure the thickness of their skin, their pH, their moisture content, their weight, their sugar content. You taste them. You take some seeds from each and determine their genetic composition. You compare that to the genetic composition of a pear from the sickly tree. With each measurement and each comparison, you learn something more about the process of growing pears. Sometimes the variation has no obvious cause, and you must look further. But the more you study the variation the more you learn about how to improve the process and how to eliminate waste.

Obviously the workers picking the pears do not determine their quality. Sure, a careless worker can cause bruises, but the number of bruised pears is most probably determined by the methods and tools used for picking and the training given to the workers. The person who controls the system, the farm manager, largely determines the quality. The same is true in factory and office operations. More than 90% of the problems are due to the system imposed by management at all levels.

CAUSES—COMMON AND SPECIAL

Dr. Deming names the causes of the variation and problems inherent in the system *common causes.* In a stable system or process, these variations are random, caused by the way the system has been set up and is managed. In a process that cuts pieces of a steel bar to the "same" length, the exact length of the next piece is unpredictable, but the range of lengths is very predictable. Narrowing the range requires changes to the system. Common causes affect all the people working in the system, but the workers generally can't control them. Common causes can only be removed or diminished by changing the system, which is the responsibility of management. Some examples of common causes are poor lighting, excess variation in temperature, poor machine maintenance, inadequate training, and defective raw materials.

One-time or non-random causes of variation are called *special causes.* Special causes come and go but are not part of the system. They include operator errors, machinery malfunctions, and external events that affect the system like a power outage or a traffic jam that causes people to be late.

Consider a complex process like collecting accounts receivable. Let's take as the fundamental variable the number of days between shipping the goods or performing the service and banking the payment. Variation inherent in the system might result from such common causes as the salespeople's lengthening the payment terms at a customer's request. Or perhaps those responsible for phoning late payers procrastinate because there is no set schedule, and they have other duties that they enjoy more.

In this case, system changes—eliminating the salespeople's authority to change terms and setting a firm schedule for calling delinquent accounts—would narrow the range of variation. They would also probably reduce the level of the variable (number of days to collect receivables). Other changes in the system that could reduce the level would include speeding up mailing of invoices and responding more promptly to customer complaints. These are all changes in the system that tend to reduce the *common* causes of variation. On the other hand a postal strike would be an example of a *special* cause of variation affecting accounts receivable. Temporary changes in the system to reduce the effects of the mail strike would be necessary, but these would not be permanent changes in the way the system is designed to work.

To make use of variation, you first need to decide what variables in a process to measure. This is an important decision, since you want to pick those fundamental variables that most affect the outcome of the process. It is not always as easy as picking variables like the length of a piece or the days receivables are outstanding. In a complex process, there may be hundreds of variables, with numerous interrelationships. You need to establish which ones are most likely to determine the outcome of the process. For example, if you are studying a process for coating carbonless copy paper, the weight coat is a fundamental variable. However, that variable may be affected by a number of others, such as machine setting, temperature of the coating, ambient temperature, coating viscosity, percent coating solids, and type of solvent used. Perhaps it will be sufficient to measure only the coating thickness and the machine setting, but you cannot be sure, initially.

To find out what variables to focus on, ask the experts **at all levels.** Ask the people running the machine. Ask the people who see the changes in the process and the effect of those changes on the end product. Sure, your engineers will have some ideas, but don't let them work in a vacuum. Make sure they understand that one of the best sources of knowledge about the problems in a process are the people operating that process. In a fast food restaurant, the order takers and the people in the kitchen will know where the problems arise. In an insurance company, clerks processing applications will be able to pinpoint many of the complexities and troubles.

SURVEYING THE EXPERTS

To find out what the experts know about the problems, survey them. Sit down with each one and talk. You can find out a great deal about problems in a process without getting into a gripe session. Have certain questions prepared but also encourage a general discussion. Or use a written survey.

Besides designing the survey to find out what you want, you need to

1. Explain the purpose of the survey,
2. Tell what you plan to do as the result of the survey,
3. Give the results to the people surveyed,
4. Do what you said you were going to do in Item 2.

Sticking with those four elements may make the survey difficult to design, but if you ignore them, your survey may be inaccurate and may damage your relationship with the experts.

If you are not prepared to do something with the results of a survey, it is obviously a waste of time. And if you want people's thoughtful cooperation, they must know what you are trying to do, and that you plan to take action based on the results. That doesn't mean you have to take every suggestion, but you should be prepared to explain to people what you can and cannot do, when, and why. If nothing comes out of a survey, not even any communication, you will have lost people's cooperation. Therefore, limit your initial objectives. People will understand that you can't work on everything at once. In surveying the experts, your main goal is to find problems, complexities, errors, and unexploited opportunities.

A survey with limited objectives helped determine the cause of erratic swings in yield in a "clean room" coating process for computer memory discs. People suspected dirt or contamination but couldn't find the source. A survey of all people working in the area resulted in many suggestions but little progress. But people kept working together, and finally a janitor said he noticed an air movement when two particular doors were opened simultaneously. Further experimentation proved this to be a major cause of the yield swings. Persistence and a simple survey of the experts solved the problem.

The other experts, besides your employees, are your customers and suppliers. Again, before surveying your customers, you should be prepared to give them the results of your survey and tell them what you plan to do about the results. Customers can explain what adds real value to a product or service and what is waste. You also discover waste by working closely with your suppliers and asking them what you can do to reduce their costs and improve quality.

Surveys provide enough information about problems and opportunities to enable you to decide what to work on first. These things usually fall into two categories. First, tackle those things that everyone agrees should be done and that require little or no effort to change. For example, if you find invoices are being mailed a day late because they usually miss the last messenger's trip to the mail room, you should be able to fix the problem easily, provided you find out what is really going on now. Or if you discover you are spending hours preparing a report that no one reads anymore, a simple decision can eliminate that waste. You might go further and improve the process that generates reports.

Also give high priority to problems whose solutions would make a substantial difference in your operations. One such problem might be the periodic contamination of a chemical cleaning bath, which causes rejection rates to soar at final inspection but remains undetected until a substantial quantity of product has been made. Or you might want to change the way you package your product for shipment, which you might discover causes 60% of customer complaints. These types of problems have greater dollar impact than the first type but are not as easy to fix.

GATHERING DATA

Ordinarily, surveys give an idea about what to work on, but you need a lot more data before you can determine your final priorities and decide how to eliminate the waste. Data gathering is a critical step, and if not done thoughtfully and methodically, it can be a big source of waste in itself. There are four major reasons for collecting data about a problem:

1. To reveal a problem,
2. To verify the existence of a problem,
3. To analyze a problem,
4. To prevent a problem.

Obviously we could substitute the words "waste" or "opportunity" for "problem." Remember, our core activity is "to identify, quantify, and eliminate waste through process improvement." Historically, much of the data gathered about any process or system has come in the form of a final inspection. Inspectors asked questions such as, Does a product meet specifications? How many "good" ones are there and how many "bad" ones? Can they be reworked?

Usually this type of inspection data reveals waste and verifies that it exists, but it does little to help analyze it and prevent it. It gathers data about results and not causes. To get at the cause of the waste, you need to go back in the process to gather data at each operation where defects may possibly arise. To do something about the defect, you need to discover what happens in the process that causes it. It is hard, time-consuming work, but if you can't discover, in great detail, what is causing your problems, it's unlikely you'll be able to fix them. That is why it's so important to put a lot of thought into your plan for gathering data.

One company that I worked with had a severe problem with imperfections in the paint on the metal surfaces of its product. Production on some days was much worse than on others, for no apparent reason. The company gathered a great deal of data at inspection stations about the number, size, type, and location of the imperfections. Management agonized over specifications, trying to decide which imperfect products could be shipped without unduly jeopardizing the company's reputation. Invariably, these criteria would be relaxed toward the end of the month, when everyone was concerned with meeting the sales budget. The inspection station data revealed very little about what was causing the problems.

Then they decided to gather data at each step of the process. First, they formed a team to identify all the probable causes of the problems. This team included production workers from each of the processing steps as well as engineers and managers. They came up with things to measure, such as surface finish, surface cleanliness, paint viscosity, timing of cleaning solution filter changes, and number of dust particles in the air. They reached a consensus as to which were the most likely causes of the problem. They carefully defined what they would measure and prepared data sheets so that operators could record the measurements. Operators were asked to fill the "comments" space on each sheet with their best guess as to what was happening when significant changes occurred. Eventually the team discovered that surface preparation and surface cleanliness were the key variables. By paying particular attention to these variables for four months, they reduced paint defects by over 90%. The important change in their procedures was gathering data about *causes* rather than *results*.

It was not an easy process. One of the keys was identifying the *fundamental variables* in the process. A team made educated guesses and then refined those guesses through trial and error. The data they gathered eventually told them the key variables that were fundamental to running a trouble-free process, with minimum waste. It was also difficult to come up with *operational definitions* of what they wanted to measure. It's one thing to say measure surface preparation and another to define exactly what you will measure and in what manner. Using scientific instruments to measure surface roughness on a production line was not practical. So instead they established a visual scale for roughness and uniformity and provided photographs to keep the operator's eye "in calibration." These aids provided a rough but practical operational definition of surface preparation. The operator preparing the surface recorded the data, and often the operator could discern a reason for a change in roughness or uniformity. Operators didn't try to examine every piece but followed a random sampling plan established with statistical principles.

This example illustrates some important principles of data gathering:

1. Create a clear definition of the problem to ensure you collect the right data.
2. Come as close as you can to measuring causes as well as results. You need both!
3. Identify those variables that are fundamental to running a trouble-free process. Ask the experts (the operators of the process) to help you.
4. Precisely formulate an operational definition of what you want to measure. Eliminate ambiguity. Specifying that a disc be made "flat" is insufficient. Something that is "perfectly flat" for one application may be "severely warped" for another. The definition needs to include what is measured and exactly how it is to be measured.
5. Carefully select the right measurement technique. Consider such factors as precision, accuracy, and ease of use. Be sure the technique is practical in terms of the cost of making the measurement.
6. If sampling is used, it must be truly random, with a sound statistical basis. (See Ishikawa, Note 9.)
7. Whenever possible the operator of the process should make and record the measurement on a real-time basis. This allows on-the-spot observation as to what is causing variation. These same principles apply to administrative processes as well.

Think about collecting accounts receivable. Some of the fundamental variables identified in one operation were

1. On-time shipment and/or delivery of the product,
2. Timeliness of invoice mailing,
3. Pricing accuracy,
4. Accuracy and completeness of shipping documents,
5. Credit terms,
6. Timing and frequency of follow-up calls, and
7. Product complaints.

The company established operational definitions that enabled it to measure these variables. It randomly sampled invoices and shipping documents. It uncovered and corrected the causes of its accounts receivable problems. Within four months, it had reduced the days of receivables outstanding by 18.

It sometimes requires a little more thought to discover how to measure the fundamental variables in an administrative, R&D, or legal process than in a manufacturing process. But they are both processes, and the same principles of data gathering apply. Organizations are not so used to thinking about waste in an administrative process or in other knowledge work, such as research and development. The waste is there, although often it is not so easy to see. In fact, because organizations have not traditionally identified and quantified administrative waste, it is usually much more prevalent than they might imagine.

Become accustomed to thinking of everything as a process. Then you can identify problems and causes, gather data on the fundamental variables, and use that data to find ways to eliminate the problems and the waste.

KEEP YOUR EYE ON THE SYSTEM

Even with accurate measurement techniques, an organization can make little use of process variation unless it looks in the right places for the causes of the variation and focuses on the truly fundamental variables. The new system of management uses the work of statisticians to change management's ideas about what causes variation in the process and how it can be controlled and changed. The coldly numerical world of statistics provides a convincing rationale for treating people with a lot more respect than they have often been given under the old system.

Most managers assume that once a process has been established, it will work the way it was designed to work unless someone who is a part of the process does something wrong. In other words they assume that variation in a process is caused by and can be controlled by the people running the process. This thinking applies to all kinds of processes, including those in manufacturing, paperwork, and services. For hundreds of years, people have thought that this assumption was just common sense. If production yield drops from 92% to 88%, the people running the process—or perhaps those who supplied the parts—must be responsible and their way of working must be changed.

Statistical evidence tells us something quite different. *It says that in a stable process, 90–95% of the variation is inherent in the system and can't be controlled by the people working in the system.* Only 5–10% of the problems are due to the people operating the system. Since management establishes and controls the system, only management can address most of the system's problems. They are problems of the system, which workers in the system can't control. Nevertheless, the workers in the system are in the best position to *identify* the problems caused by the system itself. That is why Dr. Deming calls the workers "the experts." Since they are working in the system daily and see problems when they occur, they obviously are in the best position to identify those problems. But they are not in a position to do much about the problems in most cases because they usually do not have the power to change the system. Either the managers must change the system or the organization should empower the workers to do it. Managers usually have the power, but they generally assume that workers cause the problems. Therefore many traditional management theories say that the worker must be tightly controlled and disciplined to improve performance. This advice results in the authoritative mindset of the old system. Furthermore, the old system assumes that those in authority possess the important knowledge. The workers, far from being the "experts," are simply imperfect tools required to operate the system.

Lockheed Aeronautical Systems Company effectively used its experts to combat the minor airplane body defects it calls "squawks"—scratches, tool marks, and protruding rivets. After inspection of a plane in September of 1989 revealed 405 squawks, the company decided to combat the problem with "target teams" composed of production workers, inspectors, managers, and other personnel. The teams track the airplane subassemblies, discovering persistent problems and using the combined experience and creativity of their members to brainstorm solutions. The production worker responsible for the job where the squawk occurred is not punished but treated as an important resource in the effort to eliminate the problem. Often the team discovers that improved tools or training can ensure that the problem will not recur.

Six months after the target teams were established, the squawk count had dropped to about 100, one-quarter of the former rate, and it is now below 90. Relations between Inspection and Production have improved tremendously as everyone works together to lower the squawk count even further. Supervisors give production workers much of the credit for coming up with successful ideas. And as one supervisor put it, "Our most important benefit is the pride our people have in knowing they have built quality into the [airplane] and the enthusiasm they show for continued improvement."[8]

A whole new philosophy of work evolves from understanding that 90+% of the problems in an organization can only be fixed by those working *on* the system and that those working *in* the system possess most of the knowledge of system problems. Improvements in the system should be a cooperative venture. The people working *in* the system should be asked to identify the problems which cause waste and to help the managers working *on* the system to fix the problems. Because both groups need to work together to make continuous improvements, management needs to develop a spirit of teamwork to get the cooperation of all concerned. Managers need to learn to listen to their worker "experts" and to provide them with the training necessary to make them more effective. Once the workers realize that management sincerely wants and values their help and advice, miracles can happen. Work is much more meaningful. People take pride in what they are doing and in their organization. They understand that their work is important and that they can make individual contributions. They feel appreciated by management and by their peers. After we changed to the new system at Nashua, absenteeism in the computer disc operation dropped from 7% to under 3% in just 8 months. Workers felt respected and proud of their accomplishments.

Dr. Deming is today heralded as a genius in part because he saw that Western organizations were managing entirely the wrong way. *They were working on controlling effects and not causes.* They were trying to inspect quality into a product or service, instead of insuring things were done correctly each step of the way. They were prodding people to "do a better job" instead of using their brains and energy to help identify and correct problems in the system.

Chapter 5
CHARTS

As anyone who has received a stack of computer printouts knows, data, in its raw form, is usually difficult to analyze and discuss with others. The large amount of raw data gathered about the variation in a process needs to be converted into a more useful form. Fortunately, statisticians have developed a number of useful tools to turn data into information. Chief among these tools are simple charts. A single chart communicates more than pages of numbers. Charts make the data understandable and provide information about where the waste is and how to eliminate it. They help people within the organization review the data, communicate about it with others, discover problem areas, and see progress.

THE LANGUAGE OF PROCESS IMPROVEMENT

Many business people are familiar with charts and may be skeptical about how charts can be useful tools for finding waste. What I am advocating is a whole new level of use and understanding. Everyone from the CEO to the experts operating the process needs to become bilingual, fluent in two languages—English (or their own native tongue) and charts. To get full value out of charts, the organization pursuing The Right Way to Manage must measure the right things and use the charts to identify and fix problems. Charts need to become a part of the way we think, talk, work, and act. Charts are a crucial language for process management.

Shortly after we started working with Dr. Deming, I recall a conversation I had with a vice president about his use of charts. With a touch of pride in his voice, he said something like this: "Bill, I can see why we need to put charts on our machines, so the foremen and engineers can see what is going on. Charts are good for that level, but what I use are the financial statements. Those statements 'sing' to me. Give me an operation's profit and loss statement and a calculator, and within 45 minutes, I can tell you everything you need to know."

I won't tell you what I said to him, since I was still in the old authoritative mode of management. What I should have said was: "Larry, financial statements are fine for checking the condition of the forest, but we have to start looking at smaller groves of trees and individual trees, sunlight, rainfall, soil conditions as well. Financial statements show the results and tell us when we have a significant problem, but that's like inspecting for quality after the product is finished. It's too late to prevent a lot of bad stuff from being made. Financial statements are also so far removed in time and place from the actual operations that they often obscure the causes of our problems. They show us the *effects* of our processes, but they don't show the causes. We really need to identify all the significant fundamental variables in each of our processes and chart them in such a way that we can monitor and analyze them and work to make *continuous improvement* by identifying, quantifying, and eliminating the waste. That's 'real-time' management, and not the 'after-the-fact' management that you can do with financial statements. As Dr. Deming says, just using the financial statements is like driving an automobile by looking in the rear view mirror! Real-time management charts identify problems at their source, so they can be analyzed and fixed right away. They also help make the most important decision of every day—what to work on."

Larry eventually got the picture, but not because of what I said to him at the time. He became convinced of the power of charts the same way everyone else does—by using them daily and seeing what they could do for him. No amount of study or discussion is as convincing as trying something and finding that it works and brings about dramatic improvements. Successful results are the best motivator.

Besides identifying problems on a real-time basis, charts are powerful communication devices. They quickly and dramatically show management

1. What the significant issues are,
2. What to concentrate on,

3. What is happening (trends),
4. If a problem is being fixed, and how quickly,
5. When a problem is fixed and if it is staying fixed.

Figure 5-1 on the following pages shows the seven types of charts I have found to be the most helpful.

The run chart, Pareto chart, correlation chart, histogram, and control chart are used to describe and analyze variation. The flow chart is a map of a process, and the fishbone (cause and effect) chart helps develop ideas for solving a problem or improving a process.

These seven charts are invaluable in defining and communicating problems and opportunities. They can help fix a problem and indicate the level of improvement. They can monitor a process to make sure the problem stays fixed. They are powerful tools for continuous improvement.

The summaries that follow do not attempt to describe technically how charts are constructed. Instead, they explain how charts fit in the new way of management and when they are appropriate for use. For an excellent text on charts that is suitable for self study, I recommend Ishikawa's *Guide to Quality Control.*[9]

RUN CHART

A run chart is a time plot of events that shows the level of operations, how that level varies, and if there are trends or not. Run charts usually take the format shown in Figure 5-2.

This is probably the most common and most familiar form of chart. It is used to show a time sequence of whatever is being measured—such things as production level, sales level, absenteeism, backlog of engineering hours, and days lost due to accidents. It is especially valuable in helping to visualize trends in any variable or process.

Figure 5-1 Seven Simple Charting Techniques

Pareto

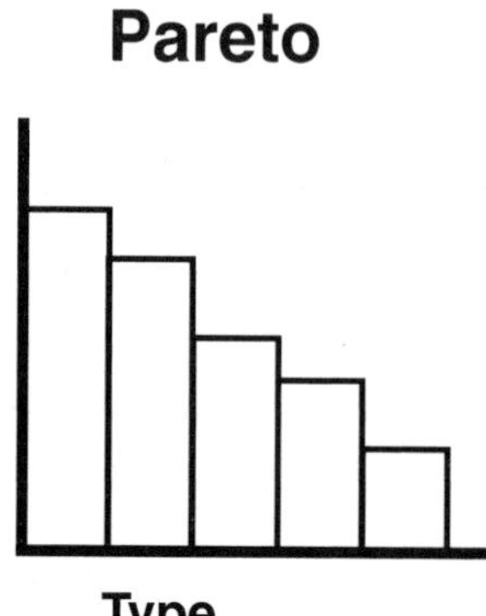

Type

Fishbone

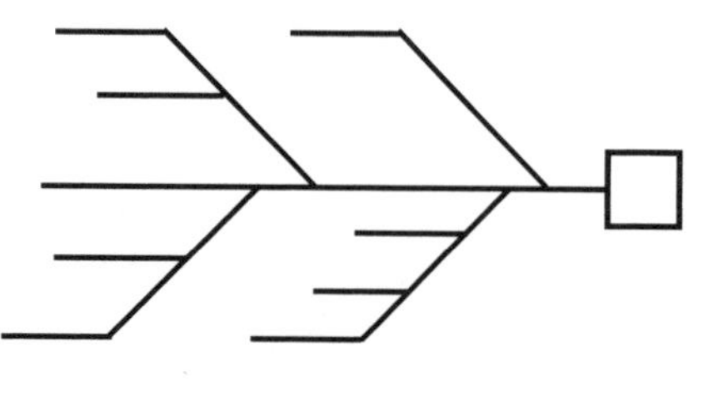

Linking Conditions to Results

Block Flow

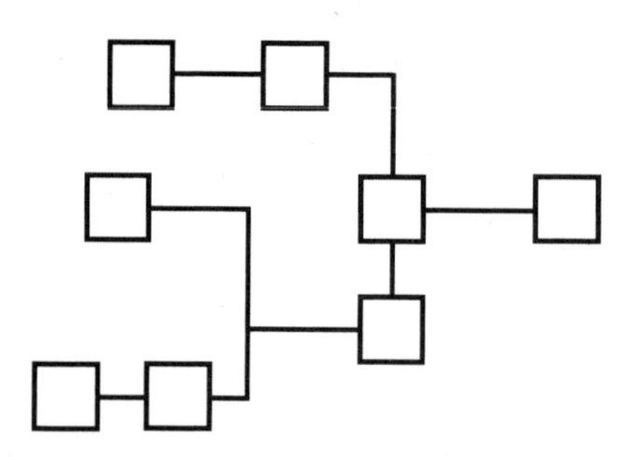

Connected Events

Run

Time Plot of Events

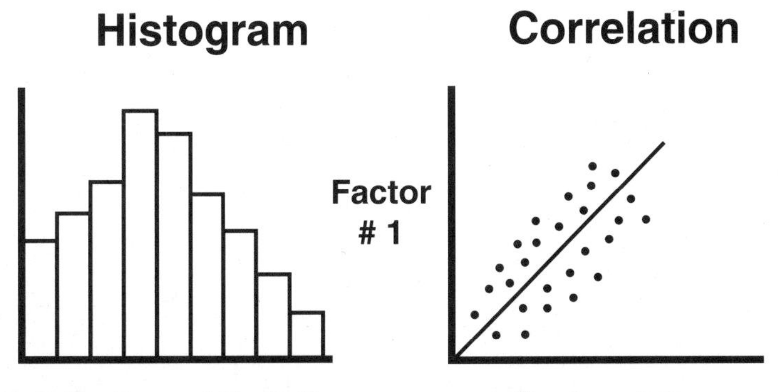
Histogram
Correlation
Factor
1
Distribution of Variation

Factor # 2

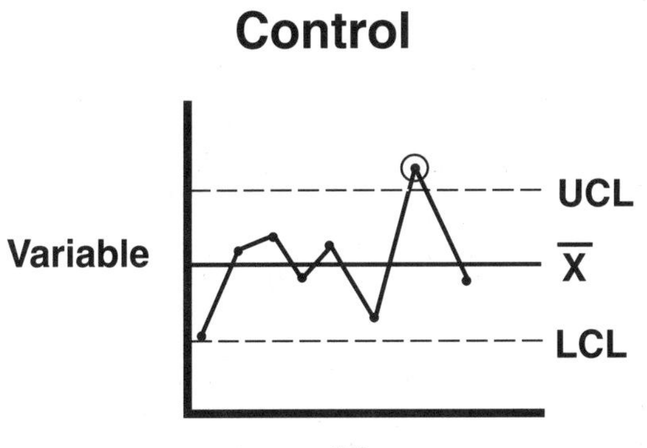
Control
Variable
UCL
$\overline{X}$
LCL
Time

Figure 5-2 Run Chart

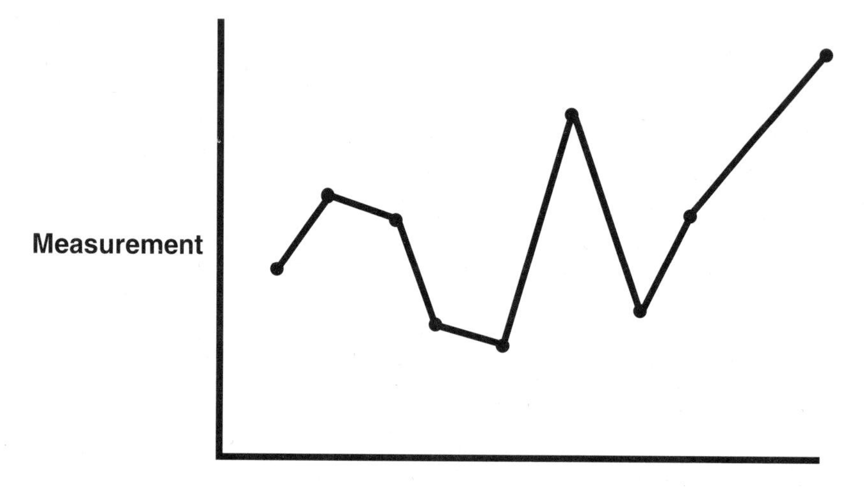

Each person responsible for a process or any part of a process should plot those variables that best describe the important things that happen in the operation. It might be useful to plot such things as the percentage of material wasted each day, the percentage of time R&D chemists work on what counts, the number of sales calls made by each salesperson each day, the days of receivables outstanding each week, the days of inventory broken down by major components, the temperature outside or in an oven. Figure 5-3 shows some charts commonly used in running a business.

Figure 5-3 Running a Business

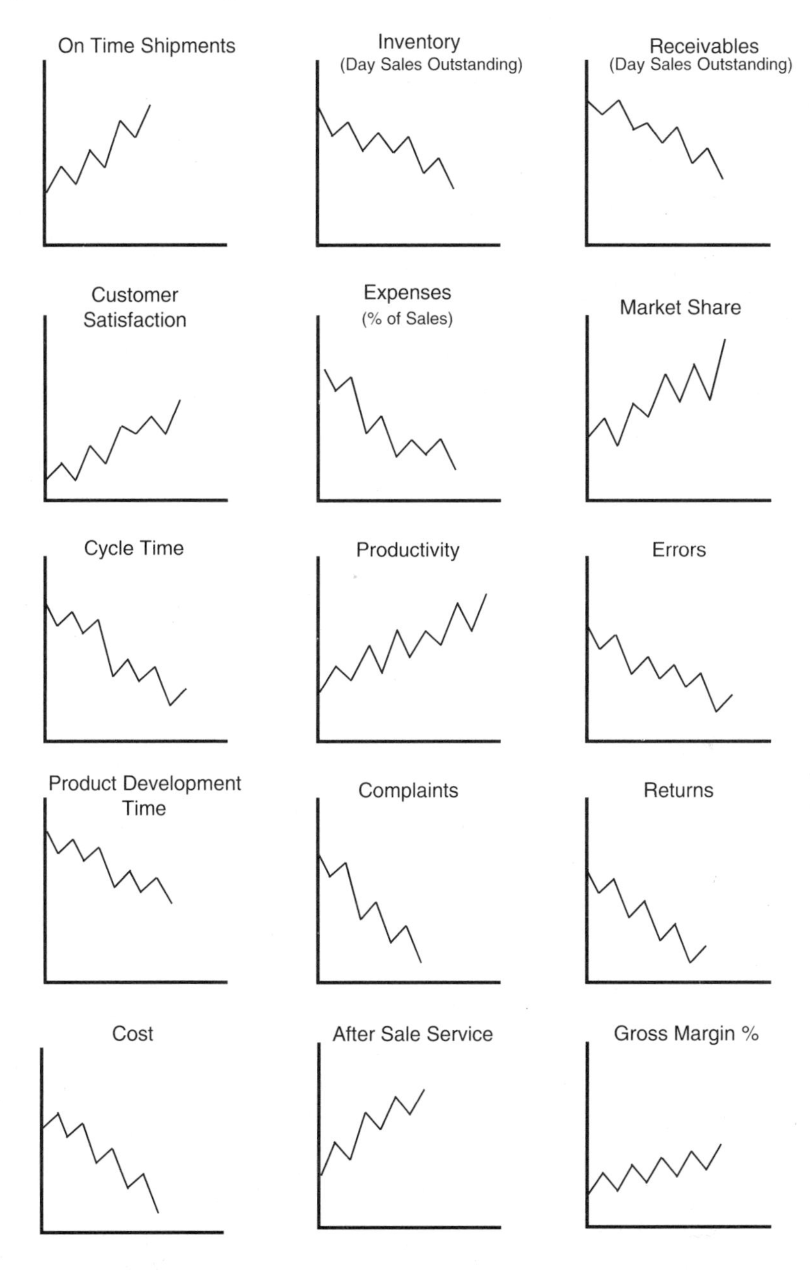

Run charts enable managers to see what the current levels of operation are and to observe the amount and level of variation in a process. They readily identify trends, favorable or unfavorable. This evidence can lead to analysis and imagineering and ultimately to new levels of operation. Without run charts, an understanding of a process may be largely subjective.

As managers use run charts to help them think about the system, they should constantly ask themselves questions such as

1. What is going on?
2. Do I like the current level? If not what should be done?
3. Has anything changed?
4. Do I like the change?
5. Is the change what I tried to do? Is it in the right direction?
6. Are things getting better? Or worse?
7. Are any trends showing up? Are they what I expected?
8. Should I translate this run chart into a statistical control chart? (to be covered later)
9. What other charts do I need?

Run charts are one of the most powerful communication devices of all. They are easy to understand. Everyone can see the level of operations and whether continuous improvement is being made.

They also communicate a message about what management is focusing on. Posting an updated run chart of a specific variable in a prominent place sends a message about what a particular manager considers important. Generally, people will respond by keeping up to date on that variable, and they will look for ways they can contribute to its improvement. Therefore it is important to chart and publicize those variables that you want to be priorities in the system of continuous improvement.

Most people are already using run charts in one form or another. The question is what run charts could or should you use to measure process improvement? Which will best reveal progress in reducing or eliminating waste? What would the run charts look like if everything were right?

HISTOGRAM

Three charts help us describe and understand the variation in any process: the run chart, the histogram, and the control chart. Each presents data on the variables in a process in a different manner. Together they tell us a great deal about a process and help us to understand, control, and continuously improve the process.

The histogram is a powerful tool that helps answer the question "How does a process vary?" It gives us a different perspective than a run chart. While a run chart shows how a process varies in time, the histogram accumulates measurements according to the frequency with which they occur. The height of the bar in a histogram reflects how frequently a given value occurs.

Figure 5-4 Histogram

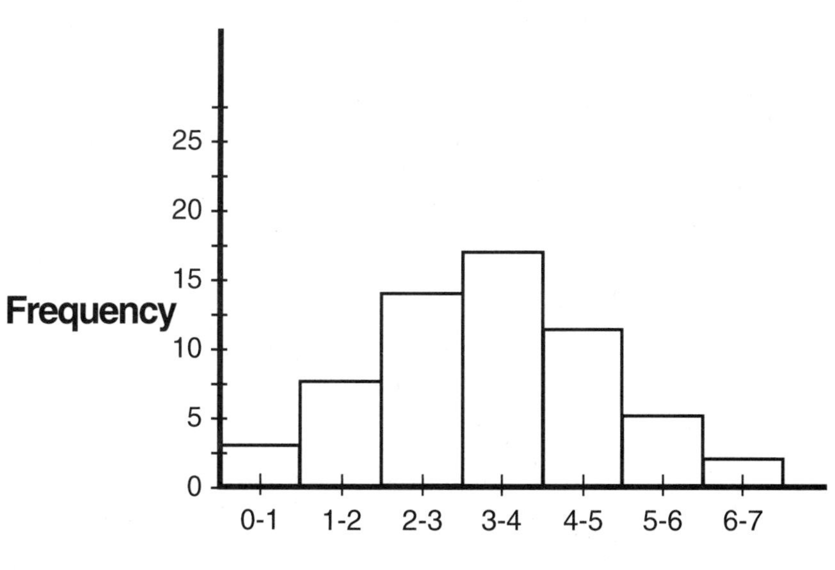

The sample shown in Figure 5-4 approximates a normal or bell-shaped curve. This curve is typical of the distribution of measurements of most natural and industrial values like the heights of women, auto door frame widths, shaft diameters, and required voltages. Measurements or values of a variable tend to cluster near the average and diminish rapidly as the values move away from the average on each side. The histogram is a powerful communication device because it provides information on both central tendency—average—and dispersion—variation.

Like the other charts, the histogram can be extremely useful for people who are not directly involved in business and who never think in terms of "quality." I once happened to fly across the country next to a special teams coach for a National Football League team who thought that my work in quality was totally unrelated to his job. I asked him what the important details of the game were for him, what would cause his team to lose a close game, and how he could improve his coaching. After studying his team's results for the past few years, he concluded that his teams needed more than anything to block more field goals and extra points. To discover how to achieve that goal, we examined the process or system of kicking a field goal by diagramming the players, focusing particularly on the center, the holder, and the kicker (Figure 5-5). By studying videotapes, the coach could time how long it took the ball to go from the center's hands to the holder and then back to the line of scrimmage after it was kicked. Using a histogram, he determined that one of his opponent's kickers took .92 seconds on average to get the ball to the scrimmage line. Other histograms helped the coach determine exactly where the kicked ball usually crossed the line of scrimmage and how high it was at that point.

The rest was easy. The coach chose a lineman to be the blocker and lined that man up in the expected path of the kicked ball. All week the lineman practiced his jumping and timing so that he would reach the top of his jump .92 seconds after the ball was centered. The histograms and practice paid off as the special teams blocked more field goals and extra points than they thought possible.

Figure 5-5 Blocking Kicks

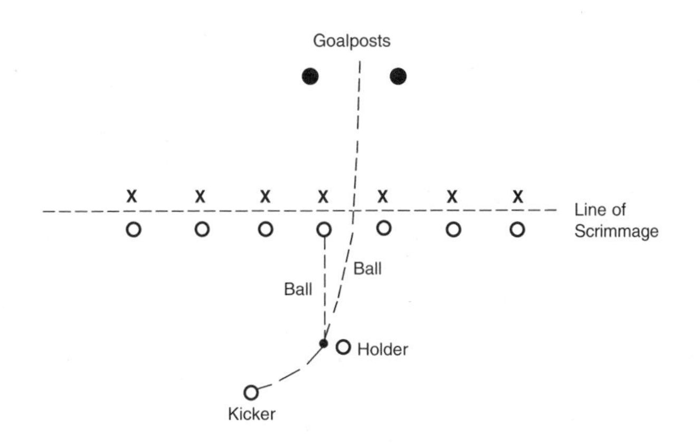

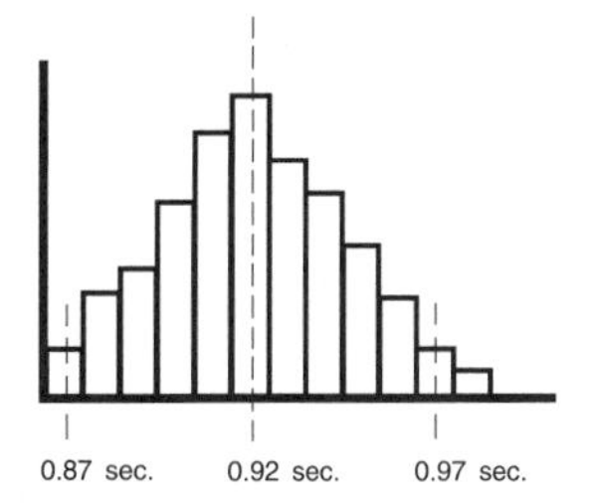

Time from when center centers the ball until ball passes over the line of scrimmage

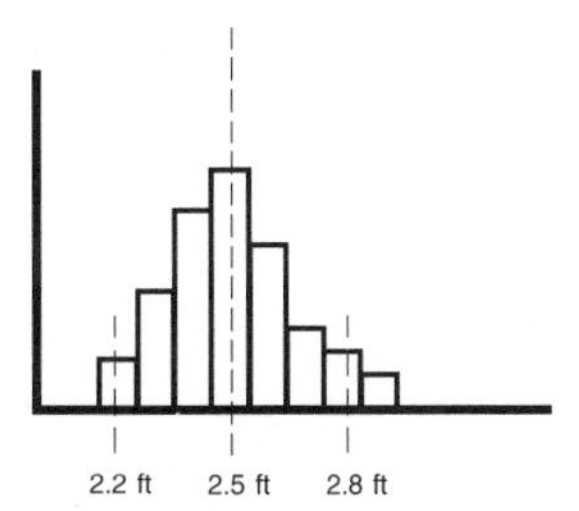

Distance ball passes to right of center when passing over line of scrimmage

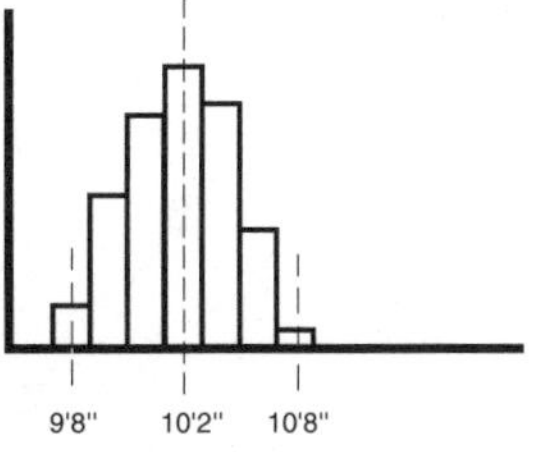

Height of ball passing over the line of scrimmage

The shape of a histogram often provides clues to the causes of problems. If the distribution does not form a normal curve, but is skewed left or right, or has two major peaks instead of one, the process needs to be examined more closely. The appearance of two peaks on a histogram like that in Figure 5-6 usually indicates that more than one process is involved. Perhaps two different machines or two different shifts have different production averages. Any significant deviation from a normal curve is a clue that something out of the ordinary is happening in the process or in the way the measurements are taken.

Figure 5-6 Histogram with Two Distributions

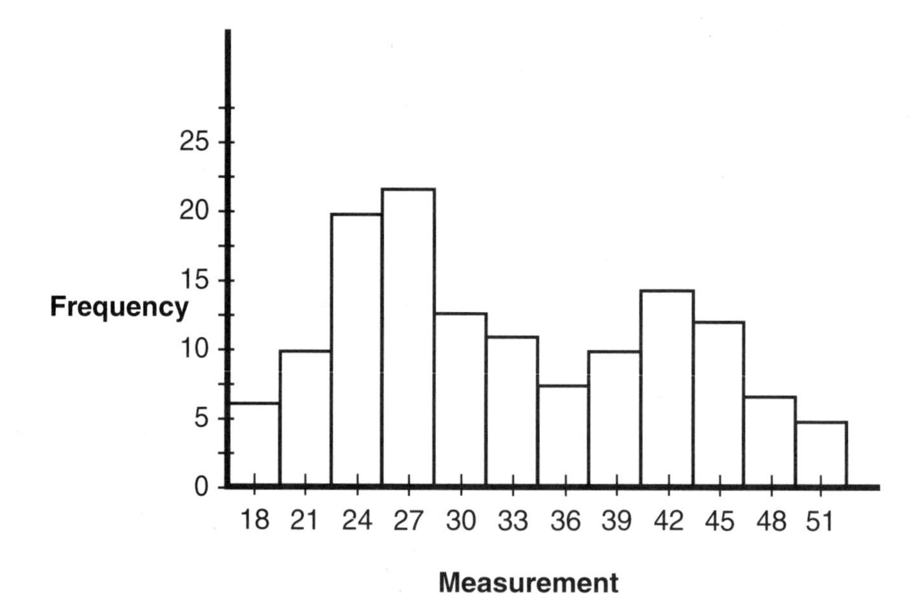

The histogram also proves valuable in describing the capability of a process. For example, if a company ships products from a distribution center to customers in, on average, three days, the company may feel that it is meeting a customer's requirement for receipt within four days. However, the actual shipping time may vary from one day to seven days; perhaps 9% of the shipments take more than four days. The customer getting the shipments after four days is not going to care what the supplier's average shipment time is. Instead the customer will be looking for a more reliable supplier who can ship within the required range.

In manufacturing processes, knowing the process capability is extremely important. If a customer wants a metal bar to be 7.5 inches +/- .01 inch, then a histogram of the lengths a supplier actually produces provides a good picture of whether the supplier can meet the requirement. If the range of values is greater than +/- .01, then obviously the process cannot meet specifications. The supplier will have to narrow the range of variation.

Historically, customers would sample a product upon acceptance to get information about purchased materials. If they were buying the metal bars just discussed, they might measure 10% of them and decide whether to accept or reject the lot. More and more, customers are realizing that this is a costly and inaccurate process. The sample provides relatively little information about the material and about the supplier's process. Both parties benefit if, instead, the supplier furnishes a histogram of the process producing the bars, or even, as is possible in some cases, a histogram of the length of bars actually being shipped. Such charts tell customers much that they need to know about the material received and also reveal a lot about the supplier's process capabilities. Assuming customers trust the supplier's honesty, they don't need an incoming inspection. More and more customers and suppliers are working this way.

CONTROL CHART

Combine the run chart and the histogram with a statistical formula, and you have the control chart. This is probably the most powerful tool of all for controlling and improving a repetitive process. It can also help further refine a description of process capability developed with a histogram.

Walter Shewhart invented the control chart in the late 1920s to improve understanding of the variation in a process. Because the control chart is based on mathematical formulae that sort random from non-random events, it separates the causes of variation into the two classes discussed in the last chapter: those that are a part of the system (common causes); and those that come from outside the system (special causes). A special cause can be identified in Figure 5–9 on page 79, because it produced a value that fell outside of the control limits.

What can a control chart do for a process?

Let's take an example from the disc coating operation at Nashua Corporation. Each hard memory disc is 100% electronically inspected for its functional performance. We had been operating at a level of 60–70% yield for several years, slightly above the industry average. One of the critical factors affecting that performance is the thickness of the magnetic coating. We had technicians in the coating room who did nothing but take production samples and measure the thickness on a very expensive piece of test equipment. If they felt it was too thick or too thin, they would adjust the coating machine accordingly. They ended up adjusting the equipment frequently. This kind of "tampering with the system," as Dr. Deming calls it, results in actually making things worse, adding more variation to the process, even though the technicians were trying to improve the system.

Dr. Deming asked us to do three things. First, let the process run without adjustment to determine the process capability. Second, make repeat measurements of the same piece to see how much the test equipment varies. Third, establish control charts for the coating thickness and train the operators to make the measurements and keep the charts updated. The operators were to stop the process when a point showed an out-of-control condition.

First we learned to our dismay that the test equipment had as much variation in its performance as the coaters themselves. Next we discovered that the process was much more stable than we had thought. It could run for a long time and perform within our thickness specifications without adjustment. The third and most surprising fact was that the operators began telling us all sorts of things that were wrong with the process, and they could relate them to variations on the charts. They had literally hundreds of suggestions, and most of them turned out to be valuable. We had always depended on our engineers to tell us what was wrong, but now we were learning from the experts, the people who were running the process. They saw things happen on a real-time basis and could link cause and effect.

The change in their attitude was eye-opening. We had never asked their opinions before but often criticized them for handling damage and other defects. Now we were listening to them and making changes based on their suggestions. They were controlling the process, and they could stop it with the guidance of the control charts. They now felt ownership of the process and pride in their work. They were part of the team that was making great strides in improving the process. Absenteeism dropped and morale and productivity soared.

The engineers also changed the way they worked. Previously they would make an experimental change in the process and look for its effect. Because of random variation, the process sometimes appeared to improve when it didn't, and vice versa. The control chart allowed them to determine accurately if they had narrowed the range of variation or changed the level of the operation. They also felt that they could now control and improve the process logically, and they teamed up with the process operators to attack process problems with enthusiasm. The yield went from 68% to 94% in less than nine months. The same operators were running the process with the same supervision and help from the same engineers, but now everyone understood the information in the variation, thanks largely to the control chart.

In our disc operation, the control chart

1. told us that our measuring equipment was out of control.
2. told us when to adjust our equipment. We reacted not to random fluctuations or inaccurate measurements but to true changes.
3. allowed us to identify special causes of variation on a real-time basis. When something went wrong, the operator stopped the process and brought in help to correct the problem before more bad discs could be made.
4. told us on a real-time basis when a process was going out of control, through either a gradual trend or a sudden change.

5. described for us the process capability. We knew it was capable of consistently making a product within the control limits of the chart. We knew when the control limits were within the product specification limits.

6. helped the operators identify and communicate common causes of variation, which management could eliminate or reduce.

7. improved engineering efficiency and enthusiasm tremendously by giving quick and accurate feedback on the results of engineering changes.

8. gave the operators "ownership" of the process, which resulted in their fully using their brains, energy, and expertise to improve the process.

9. made everyone involved a part of the same team with a consequent soaring of morale and productivity.

Obviously we needed a management system supporting the control chart, but the chart itself was a key ingredient in changing attitudes and revealing the information that the process variation contained. "The control chart is the process talking to us."[10] It is most effective with highly repetitive operations.

How does a control chart work?

Control charts work because variation in a stable process almost always follows a normal distribution, with more measurements near the average and fewer further from the average.

Figure 5-7 Control Chart of a Process in Control

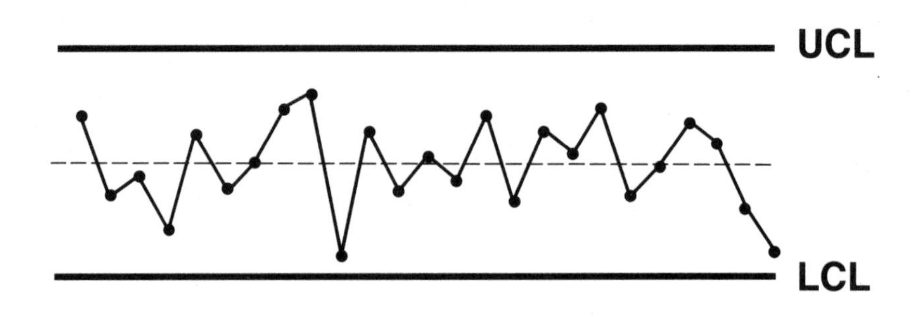

The control limits are set so that almost always in a normal distribution (99.7% of the time) random variations will fall within those limits (three standard deviations). This allows us to separate random (common) causes, which are part of the system, from special causes. The absence of special causes (points outside the control limits) indicates the process is probably stable. (A number of other tests can determine if the process is out of control even if all points are within the control limits.)

The control chart (Figure 5-7) also shows the range within which the process normally works, allowing operators to understand the process capability. It shows whether the system is capable of meeting specifications. It is important not to confuse control limits and specification limits. The control limits show what the process is producing, while the specifications define the desired limits. When engineers make a change in a process and then recompute the average value and range of values, they can see how the change affected process capabilities. The control chart is obviously very valuable as a tool in continuous improvement.

Stabilizing a process with a control chart.

Unfortunately, most processes in the typical organization are not stable, but are affected by numerous external, special causes of variation. Stabilizing a process is a significant achievement, because people can determine the capabilities of a stable process and can count on the output of the process being within a certain range. Furthermore, stability provides a base from which the effects of changes in the system can be measured accurately. This stability is critical so that people can separate the random variation from the change in level caused by the trial improvement. Otherwise they might be misled by random variation to conclude that the changes they made were either better or worse than they really were. A variable in a process which is not stable might look something like Figure 5-8 on the next page.

Figure 5-8 An Unstable Process

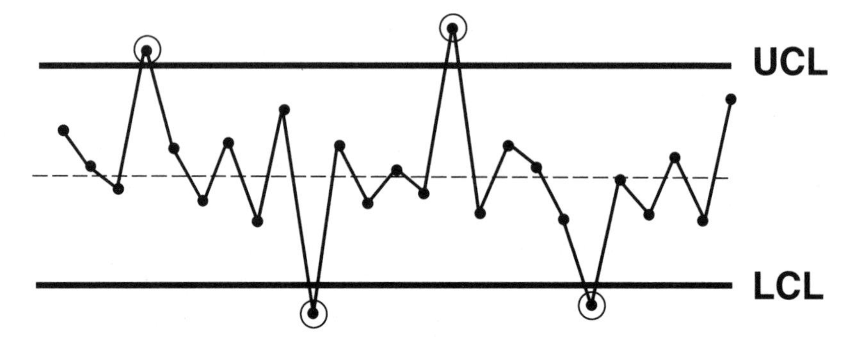

To achieve a stable process you must investigate each point outside the control limits and establish the cause(s) of its deviance. Update charts continually and investigate as soon as a point falls outside the limits. You can most easily discover a reason for a special cause as soon as it happens, not several hours or days later. Identifying and eliminating the special causes requires a lot of hard, disciplined work, because some of the special causes are not obvious. You need to be a detective gathering evidence from all the people involved, plus any available physical evidence. You need to act quickly while what happened is still fresh in people's minds and while any physical evidence is still available. If you leave too many mystery causes unsolved, you won't get a stable process.

Once the process is stable, the control chart will look something like the one in Figure 5-9, and may exhibit an occasional special cause which will require investigation. The center line of the chart represents the average value of the variable and the upper and lower limit lines represent the expected range of the data.

Figure 5-9 A Stable Process with one special cause

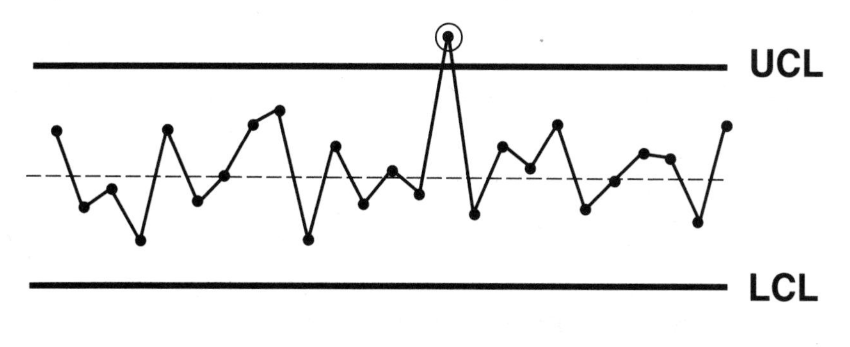

Using control charts in a stable process.

In a stable process, the control chart has two important functions:

1. to aid the continuous improvement of the process;
2. to maintain the process in control. The key people in improving the process, of course, are the experts—the people who operate the process.

Control charts are useful either when an organization wants to reduce the variation to a bare minimum or when it wants to change the level of the process *and* reduce the variation. Think back to the company cutting a metal bar to a certain length. The process may already be within the specification limits, in which case the control limits will be inside the desired specification limits, as they are in the left chart, (Figure 5-10). The company's objective then may be simply to continue to discover and remove causes of variation in order to make the bar length as uniform as possible. If, on the other hand, the range of variation shown by the control chart exceeds the specification limits, as in the right-hand chart, a certain portion of the process output will not be satisfactory and will have to be sorted out by inspection. In this case, it is more urgent to find and eliminate common causes of variation so that the range can be reduced to within the specification limits.

Figure 5-10 Process Capability and Specification Limits

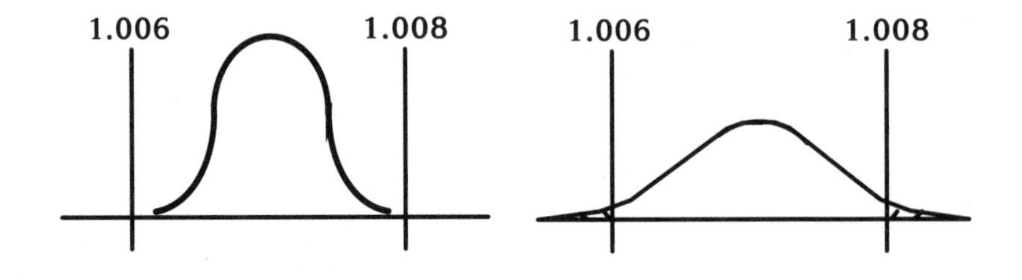

Whereas in both cases with the metal bar the goal was simply to reduce variation, improvement in most work processes means changing the level of the variable as well as reducing its range of variation. Examples are days of accounts receivables outstanding, absenteeism rate, number of grievances per month, sales per salesperson, number of burst packages per thousand, number of days from order to delivery, errors per hundred invoices, and time to develop a new product. In all of these cases the organization wants either to raise or to lower the average of the variable. If the process is stable, a managerial breakthrough is needed to make a significant change. If you don't like the way things are now, you must do something to change the process or the system within which the process operates.

Again, don't blame the people operating the process. Fix the process and the product or service will turn out O.K. Managers who work *on* the system must listen to the people working *in* the system to discover causes of waste and then take action to eliminate them. Working as a team, managers and people operating the process can make dramatic improvements. They can narrow the range, use experiments, change the level, and strive for a management breakthrough as shown in Figure 5-11).

Figure 5-11 Managerial Breakthrough

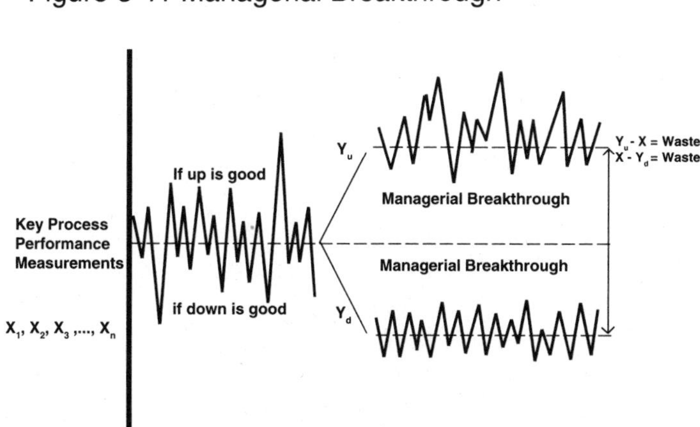

Work on the right things
Do the right things right
Do the right things right most of the time

With control charts as a major tool, the keys to using the information in variation are:

1. Ask the people operating the process for their help in identifying problems and opportunities for improvement.
2. Be prepared to react quickly to a signal of unusual variation (special cause).
3. Look for ways to reduce random variation (common causes).
4. Try experimental changes to see if you can improve the level of the process. Use the charts to identify the random variation and measure the change in level.
5. Continue to use control charts maintained by operators working in the process to identify special causes of variation on a real-time basis and to pinpoint the problems and opportunities that are common causes.

Again, keep in mind the importance of having the control charts kept by the operators of the process, rather than by technicians or supervisors. They don't necessarily have to understand how to construct a control chart, but they should be trained to plot points and interpret the charts so they can spot unusual events. Being closest to the operation, they are most likely to be able to connect cause and effect when something happens in the operation. The control chart helps them visualize what is happening, and the signals it gives help them isolate causes of variation. To be truly effective, an operator should have the authority to take some action or to stop production whenever a problem arises. Japanese companies often give their workers this authority, sometimes using a line stop indicator called an *andon* which shows a red, yellow, or green light. The green light means production is normal, while a yellow light means the operator wants some help in making an adjustment. A red light means the operator has stopped production because of a problem, most likely a special cause. This stopping of the line tends to focus effort immediately on discovering and eliminating the cause of the problem, and it avoids producing more bad product. Besides, as we have discussed, asking the operator to monitor and control the process and help in solving problems gives the operator "ownership" of the process and makes the operator a solid member of a team devoted to continuous improvement.

A stable process, at the proper level, with a range of variation within specifications, steadily and predictably yields acceptable product, making quality "control" inspections unnecessary. If problems are identified and corrected when they happen, the system doesn't produce bad product that must be further processed, sorted, and inspected. This system also eliminates practically all rework. Just imagine the cost savings from the elimination of inspections and rework!

Of course for your process to operate with stability you need a stable source of materials from your supplier(s). That is why it is important to work with suppliers who have gone through the same process of eliminating waste through tracking down process variation and who can furnish evidence of statistical control of a stable process. This evidence can also help you avoid the expense of incoming inspection.

As you benefit from your suppliers' controls, your customers benefit from your controls. The quality and uniformity of your product or service go a long way toward the ultimate objective of pleasing the customer.

PARETO CHART

Use of Pareto charts represents a way of thinking, a "Pareto mentality." Named after an Italian economist who studied the uneven distribution of wealth in the early 1900's, they are vital in making that all-important daily decision, "What should I work on?" Pareto and his partner Lorenzo showed that a large percentage of wealth was concentrated in a small proportion of the population. Similarly, they used crime statistics to demonstrate that a relatively small number of criminals committed a large percentage of the crimes.

A Pareto chart is a simple bar graph in the form shown in Figure 5-12, ranking the causes, sources, or reasons for variation in order of importance. Its function is to help identify and communicate the relative importance of problems and opportunities.

An organization might want to chart *problems* like causes of defective items produced or absenteeism ranked by employee group. Or it might want to chart *opportunities* like the various new product features most requested in a customer survey. This data

Figure 5-12 Pareto Chart

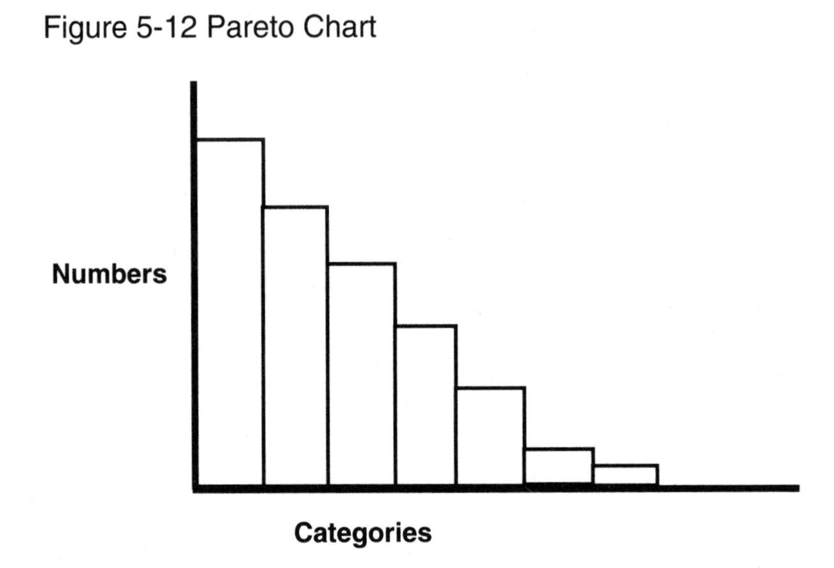

may already be available, but a Pareto chart turns the data into information that is readily understood and communicated. An organization may be producing defective units for so many different reasons that it doesn't know where to begin to improve its quality. By ranking these causes, it may find that just three or four variables produce 80% of the problems. If the top cause is creating 40% of the problems, the organization can probably reduce that cause by half more easily than it can eliminate completely five or six minor causes.

Managers sometimes think that the Pareto mentality is only for people directly concerned with production. But the higher up in an organization you are, the more important it is to have the Pareto mentality. The greater your power and freedom to act, the more important it is that you work on the right things. Top managers need Pareto thinking most of all. Only by working on the important things—the things that count—can the new management system succeed. If your major problem is lack of sales, it helps to rank the activities that can increase sales and then focus on the important ones. (Customer surveys and lost order reports from salespeople can help quantify this one.)

Whenever possible, measure in monetary terms the size or importance of a cause or source of a problem or opportunity. The number of defectives from various causes may not be the best guide for what to work on if some defectives are much more costly to fix than others or if some have particularly strong effects on customers' perceptions. In this case, multiply the number of each type of defective by the cost of that type and redraw the Pareto chart accordingly. A good way to describe this approach is a "Pareto bottom-line mentality."

A relatively small number of causes or variables accounts for a good deal of waste or missed opportunities in business. The author and consultant, Dr. Joseph Juran, coined the term "Pareto Principle" to describe the uneven distribution of causes, and he particularly applied it to management of quality. The Pareto Principle sorts the vital few from the significant many. This principle has generated the "80–20" rule of thumb: in most cases 80% of your problems or opportunities come from 20% of your sources. For example, 80% of the contributions to a church will come from 20% of the parishioners. Or 80% of a company's absenteeism may be caused by 20% of its employees.

Consistent use of Pareto charts will result in the "Pareto mentality," a persistent searching for the vital few in order to get at the important causes of waste. The successful problem-solver knows that success comes only by ranking problems in order of importance and then looking at the reasons behind the problems by breaking each into bite-size pieces. A Pareto chart not only identifies which problems are most important, it also can help determine which will take the least time to fix or require the least capital investment and still have a significant impact.

For more complex problems or processes, you may need "layered" Pareto charts. By using a Pareto chart we established that surface imperfections were the major reason computer memory discs were being rejected. "Surface imperfections" was too general a category to work on, so we made a sub-Pareto of the causes of surface imperfections: scratches, dirt inclusions, streaks, chips, rough edges, cracks, handling damage, and other. Some of these, such as rough edges, we began to work on immediately. For others, such as scratches, dirt, and handling damage, we identified further causes, and made sub-sub-Pareto charts.

How far you break the subject down depends on the complexity and the importance of the problem. A complex problem costing a lot of money deserves a very detailed search for, and ranking of, causes. Breaking a complex problem down into bite-sized pieces with the aid of Pareto charts makes it much easier to attack and focus on the most important causes first. There is one significant exception to working on the most important causes first. If, while studying causes, sources, or reasons, you discover a minor problem that can be fixed with very little or no effort, fix it immediately.

While Pareto charts help us to analyze problems and communicate about them, their greatest value is in introducing the Pareto Principle into our everyday thinking. They lead us to rank the reasons that a problem is happening or, in the case of an opportunity, rank the actions that could make it happen. They show us improvements that can be implemented immediately, with little effort, allowing us to concentrate on the major reasons, causes or sources, reduce them to small pieces, and attack them, one by one.

CORRELATION CHART

A correlation chart (also called a scatter diagram) plots one variable against another to determine the relationship, if any, between the two variables.

Figure 5-13 Correlation Chart

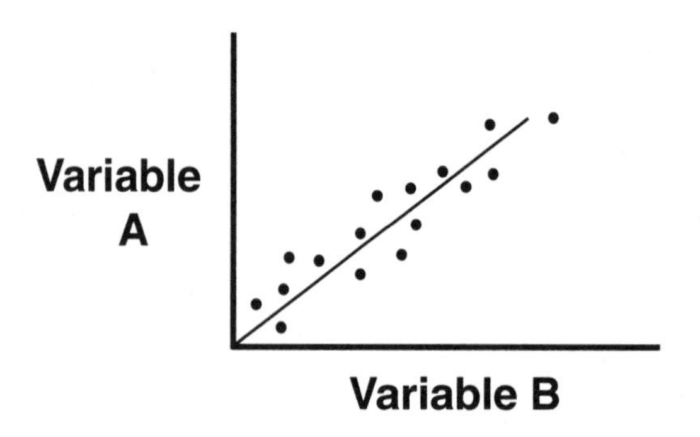

This chart is often used to show cause and effect or to indicate how one variable changes with another. For example the weights of a number of individuals plotted against their heights will probably show a positive correlation. That is, on the average, the taller a person is, the more he or she weighs. On the other hand, the weights of a sample of cars would have a negative correlation with their fuel efficiency: the heavier the car, the lower its miles per gallon.

Figure 5-14 Positive and Negative Correlation

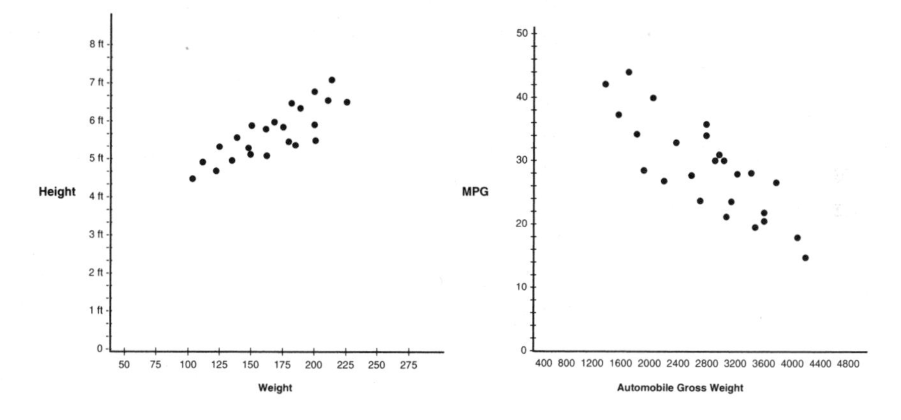

These charts can be very useful in establishing the cause of problems (or opportunities). For example, if the percent of waste for a particular product has a strong positive correlation with the relative humidity in the plant, high humidity may well be causing defective product. Or if the number of grievances filed in a week (or month) has a negative correlation with the productivity for that period, you may surmise that the workers' attitude is adversely affecting productivity—or that workers feel dissatisfied with their jobs when they're not productive.

As this last example indicates, a correlation between two variables does not necessarily establish a cause and effect relationship. If a company finds that it takes longer to collect on accounts receivable for products of lower profitability, a third variable may be affecting both collection time and profitability. For example, high waste could be causing low profitability and many partial shipments, and the partial shipments could lead to customer complaints, which lengthen the collection period.

The correlation chart is a good tool for discovering fundamental variables and checking relationships. You must also use common sense, however. You may feel that every time you plan a picnic, it rains, but common sense would tell you there is no cause-and-effect correlation between the two events. If you discover that the percentage of defective material coming from a machine correlates with the number of hours per day the machine is operated, there may be a cause-and-effect relationship, but it is not obvious. You would need to find another factor—perhaps the temperature at which the machine operates—that is connected to both of these variables.

There are differing degrees of correlation, as shown in the charts in Figure 5-15. A strong correlation is a good indication of a definite cause and effect (although sometimes the cause may be a third variable).

A correlation chart, like a control chart, is a detective's investigative tool. It can be used to check theories. Particularly when a complex process involves many variables, a correlation chart can be useful for checking the variables against each other to see which ones are interrelated.

In Nashua's computer disc operation a check of operating variables showed a strong correlation between waste due to foreign particles and the particle count of dust in the air. Although this correlation was anticipated, its strength led to redoubled efforts to provide better clean-room conditions. With the use of the correlation chart, we could easily see what money we could afford to spend to improve air cleanliness.

FLOW CHART

Flow charts are a key tool in understanding and improving any work process. They provide a map of a process that shows the way things *now* happen, or how we *think* they happen, or how we believe they *should* happen. Flow charts identify and classify each of the steps in a process and indicate their sequence and the time required for each. Historically they have been used primarily to plan a work process. In the new system

Figure 5-15 Types of Correlation

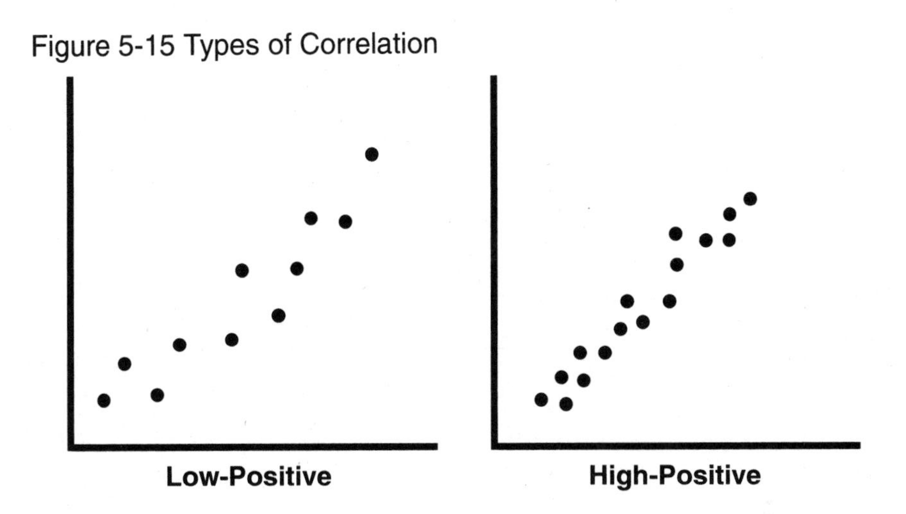

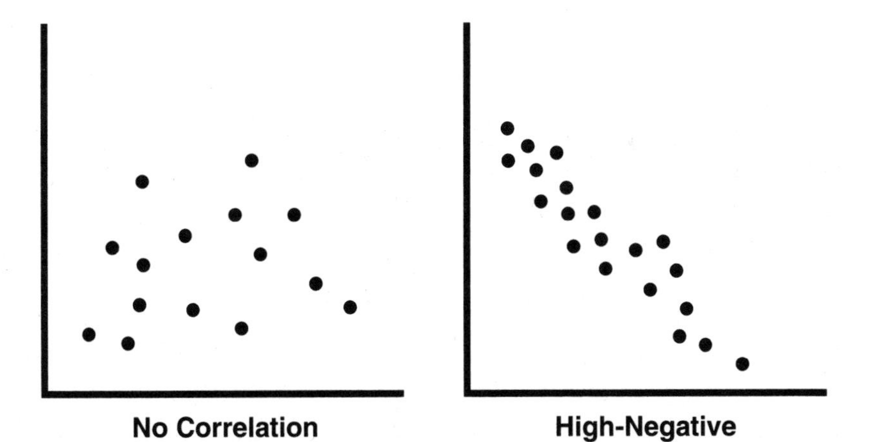

they are used as a way to find the waste in a process and to develop ideas for improvement.

Flow charts can be useful in studying things as complex as designing and building a nuclear power plant or as simple as filling your car's gas tank. Flow charts can display varying degrees of detail. An overall block diagram showing the major elements in a process often provides a start. The most common flow charts are more detailed, showing each step of the process and a time for that step. For a very detailed analysis of a process, each step can be broken down into sub-steps and sub-sub-steps, etc. Such detailed charts may be desirable for understanding and analyzing complex processes and/or processes that are identified as important contributors to waste.

A chart or map of how things are being done needs to be accurate so that you can identify the waste. Don't put it down the way you think it works. Question people to see how it is actually being done. Determine how long each step actually takes, the time between steps, and the total elapsed time. Flow charts help people visualize what is happening, and they are great communication tools for discussing process problems and opportunities. They also visually highlight non-value-added activities such as rework, inspection delays, and duplications. When you see the symbol for waiting time you know right away that the process is not as efficient as it could be in using time to add value for the customer.

Flow charts use six conventional symbols:

SYMBOL	DENOTES	EXAMPLES
○	OPERATION	Typing, drilling, soldering, filling in forms.
□	INSPECTION	Checking dimensions, reviewing invoices, matching color.
⇨	TRANSPORT	Moving material, walking with papers.
◇	DECISION	Determining if the part is good, what measurement to make, what pricing schedule applies.

 DELAY — Waiting for work, waiting for instructions, standing in line at a copier, waiting for a signature.

 FILING/STORAGE — Filing forms, entering on computer disc.

Using these symbols, setting up a flow chart can be very straightforward. It does, however, force you to think hard about and investigate how things are done.

We said earlier that you could construct a flow chart about everything from building a nuclear power plant to pumping gas at a filling station. Let's examine how these would look.

Let's assume you have the job of designing and building a nuclear power plant for a submarine for the U.S. Navy. Since this is a complex job, the initial flow chart will not be very detailed. Nevertheless it can be very helpful in understanding what work must be done.

○ Prepare design proposal (3 months)
□ Discuss design concepts with Naval Reactors Branch, Bureau of Ships (1 mo.)
D Obtain approval of design concepts (1 mo.)
◇ Appoint Project Manager (1 week)
◇ Appoint Electrical, Nuclear, and Mechanical Engineering and Budget Managers (1 wk.)
○ Develop project budget and schedule (4 mos.)
□ Submit budget and schedule to Navy and obtain approval (2 mos.)
○ Recruit necessary personnel (3 mos.)
○ Develop designs for all electrical and mechanical systems (5 mos.)
○ Develop specifications for all components (6 mos.)
○ Solicit and award bids for manufacture of all components (3 mos.)
◇ Develop schedule for component delivery with shipyard (2 wks.)
○ Expedite component delivery and resolve design problems (13 mos.)

□ Monitor plant installation and test (6 mos.)
◇ Disband project team or obtain new assignment (1 mo.)

Although this is obviously a very abbreviated form of all the activities that will take place, it does show that a flow chart can make clear even the most complex process. It can also point out things you might do differently. For example, you should probably have a preliminary discussion with the shipyard about the schedule before going out for bids on components so you will know which ones are needed first. Now that you know how to design and build a nuclear power plant, let's look at the simple operation of filling your gas tank. We will make this one even easier by using a preprinted form (Figure 5-16).

You are probably surprised that the simple act of buying gasoline has 26 separate steps. Do you see any waste in this process? To find the waste, break the process into bite-sized pieces so that you can understand exactly what is happening. This is particularly true of repetitive processes that consume a lot of people's time.

Once you have charted a process, you should discuss the chart with some of the people operating the process so that you can agree on what is actually being done. Often people working in the same process operate quite differently. You should strive to identify the optimum method of operation, standardize the process, and train each of the people to perform it in the standard manner. Of course, in continuous improvement, the people will immediately be looking for further process improvements. Just examining the way things are being done will probably reveal some obvious waste. You should also determine the amount of time spent on each of the types of activities—operations, inspections, delays, etc. Putting these in the form of a Pareto chart will provide a good picture of some of the activities that are not actually adding value during the process.

A series of questions sparked by flow charts can guide your thinking in further improving the process. Can you eliminate, change, rearrange, combine, or simplify to improve the process?

Figure 5-16

☒ Present Method
☐ Proposed Method

PROCESS FLOW CHART
DATA COLLECTION

SUBJECT CHARTED *Purchase Gasoline at self-serve*

DEPARTMENT ____________

DATE *12/25/89*
CHART BY *MJC*
CHART NO. *1*
SHEET NO. *1* OF *1*

TIME IN SECONDS	CHART SYMBOLS	PROCESS DESCRIPTION
10	● ⇨ □ ◇ D ▽	Drive car into station
10	○ ⇨ □ ◆ D ▽	Decide which pump to use
5	○ ➡ □ ◇ D ▽	Drive car to pump selected
2	● ⇨ □ ◇ D ▽	stop car at pump
6	○ ➡ □ ◇ D ▽	exit from car
4	○ ⇨ □ ◆ D ▽	determine if pay first or pump first
10	○ ➡ □ ◇ D ▽	walk to pump
2	● ⇨ □ ◇ D ▽	Remove hose from pump
3	● ⇨ □ ◇ D ▽	turn on pump
10	○ ➡ □ ◇ D ▽	walk to car with hose
5	● ⇨ □ ◇ D ▽	Remove gas cap
180	● ⇨ □ ◇ D ▽	pump gas
10	● ⇨ □ ◇ D ▽	complete pumping, Remove hose, Replace cap
10	○ ➡ □ ◇ D ▽	walk to pump
5	● ⇨ □ ◇ D ▽	shut off pump replace hose
30	○ ➡ □ ◇ D ▽	walk to cashier
40	○ ⇨ □ ◇ ◗ ▽	wait in line
5	● ⇨ □ ◇ D ▽	Advise cashier pump # & $ AMT.
2	● ⇨ □ ◇ D ▽	pay cashier
15	○ ⇨ □ ◇ ◗ ▽	wait for change
10	● ⇨ □ ◇ D ▽	count change
30	○ ➡ □ ◇ D ▽	return to car
5	● ⇨ □ ◇ D ▽	enter car
10	○ ⇨ □ ◇ ◗ ▽	search for keys
2	● ⇨ □ ◇ D ▽	find keys start car
15	○ ➡ □ ◇ D ▽	drive to exit
TOTAL 436	○ ⇨ □ ◇ D ▽	– END PUMP first –
	○ ⇨ □ ◇ D ▽	
	○ ⇨ □ ◇ D ▽	
	○ ⇨ □ ◇ D ▽	

The following are examples of specific questions to ask.

Eliminate

The first step in work improvement is elimination. Don't improve what you shouldn't be doing in the first place.

Is the operation necessary? Does it cost too much? Does the customer (internal or external) for that operation want or need it?

Is there any duplication in the process?

How can you eliminate the delays?

Can you reduce or eliminate the inspection steps?

Do you really need the filing steps?

Are some operations required to rework errors? From a previous operation? From this process?

Change

How can you change the operation to improve it?

Should you change the technology? The method? The equipment?

Can you use less costly material or service?

Can you reduce the frequency of the service?

Can you reduce the number of people receiving this service?

Can you reduce the time it takes?

Will changing the process requirements make this or a subsequent process easier or less costly to perform?

Rearrange

Is the layout the most efficient?

Can you eliminate some of the transport steps?

Is the sequence of operations the most efficient?

Should you move operations to or from your department or business?

Combine

Can any of the operations be combined? With your suppliers' operations? With your customers'? (Consider both internal and external suppliers and customers.)

Can any of the operations be subdivided and added to other operations to reduce costs?

Simplify

What would be the simplest way to accomplish the objectives of this process?

Are instructions easy to understand?

There is a good reason for asking the simplification questions last, because you don't want to simplify something that can be eliminated altogether or can be done better by someone else.

While asking all these questions, keep two elements in mind—time and the perfect process. Because the use of time (people's and machines') is usually the major cost of a process, put a time element on each step of the process to help quantify the waste.

Thinking of how the process would be if everything were perfect helps you discover the waste. It keeps you asking why it has to be done this way and sparks your creativity to come up with improved solutions. Drawing a flow chart of the perfect process and looking at the differences from the present process often sparks ideas.

FISHBONE CHART

The fishbone chart (Figure 5-17) is a powerful problem-solving tool that helps to identify possible causes of variation and other problems and opportunities. The diagram in Figure 5-18 shows why it is known as a fishbone chart. It is also called a cause-and-effect diagram since it links causes to results.

Figure 5-17 Fishbone (Cause and Effect)

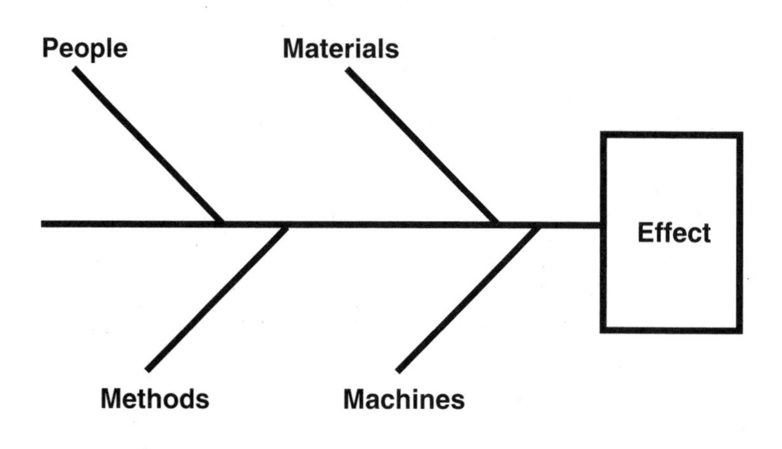

The causes can be grouped into any classifications that seem logical. If none comes to mind, the groupings of *people, materials, methods,* and *machines* can be used for almost any problem.

The following simple fishbone chart (Figure 5-18) is a compilation of possible reasons for office coffee complaints.

Figure 5-18 Fishbone Chart

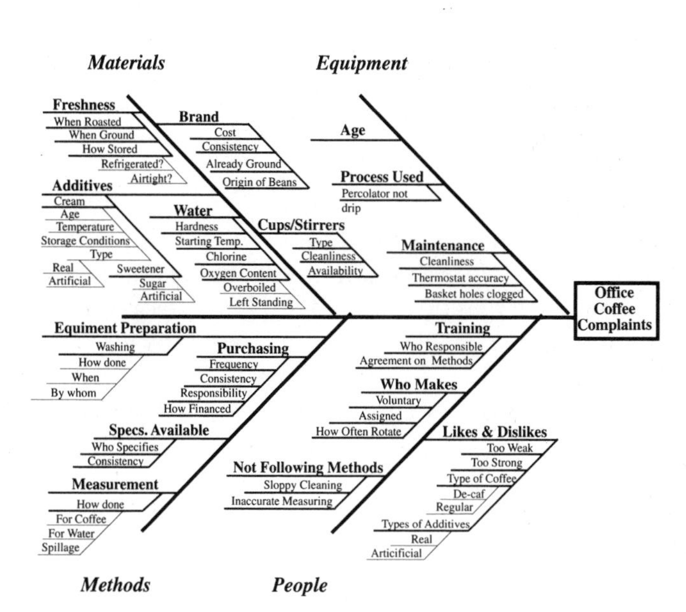

The most valuable use of the fishbone chart is to get a number of people knowledgeable about the problem to brainstorm ideas about the problem, set priorities, and develop an action plan. This method can work on any problem, from how to eliminate a particular type of defect in a manufactured part, to how to speed up product development, to how to get more sales for the company. Typically an organization might gather a group of six to ten people who are familiar with the problem and who might have some ideas about it.

It is important to have people from different areas to provide different perspectives and knowledge. For example, the group might include a salesperson, an engineer, an accountant, a production worker, a general manager, and a marketing expert. If a general manager or similar person is involved, he or she should listen most of the time and not dominate the process, or the free flow of ideas will be restricted.

First the group must agree on exactly what problem or effect to discuss. The definition should be very clear so that everyone is talking about the same thing. In many cases a flow chart of the process in question can be quite useful in developing a definition of the problem and can also serve to generate ideas about the causes.

Next everyone offers opinions as to the factors that might cause the problem or opportunity. All suggestions are listed. Since they are only opinions, none can be considered wrong. The session leader may ask for clarification of an idea, but everyone withholds criticism. An open, non-judgmental atmosphere is important to keep the ideas flowing freely. Every idea is acceptable in a brainstorming session. Even if an idea is not realistic, it may lead to others that are. At this point, the session should be fast-moving, with ideas jotted down rapidly to maintain momentum. After all the ideas have been listed, the group chooses categories for the causes (such as people, material, methods, policies, practices, environment, and machines) and constructs a fishbone, adding other ideas whenever they come up.

After completing the fishbone and discussing the causes and their classification, the group should set priorities. One of the most effective ways of doing this is to have the participants vote on the four, five, or six causes that they feel are most important. For these causes dig more deeply. Ask why they occur and when you get an answer ask why again. Ask why five times. This way you will get deeper and deeper into the process. In answering why five times you will understand the process better and/or discover where you need more data. After the results are compiled, the group can develop an action plan for gathering data on each of the major causes with someone appointed to be responsible for coordinating the data-gathering for each separate cause. A timetable should be set for the initial data gathering and subsequent preparation of another action plan. With sufficient data a Pareto chart of the causes will aid in a reordering of priorities, if needed.

This process of problem-solving has several advantages. It

- provides logical organization for the consideration of a problem;
- enables people to get ideas quickly from the group most likely to understand the problem and contribute to its solution;
- identifies many potential causes of a problem and shows the relationships among causes and effects;
- helps to categorize and set priorities for the causes of the problem;
- serves as a communication device so that everyone working on a problem or opportunity sees it in the same manner;
- makes easier communicating the problem to others who may be only indirectly involved;
- develops a sense of teamwork among the people most knowledgeable about the problem and starts them all working towards its solution in a cooperative manner.

Dr. Deming's observation about why Western managers have not followed in Japan's footsteps was, "They don't know where to look. They don't even know what questions to ask."[11] Using charts to analyze process flow greatly enhances your knowledge of what is happening. It gives you the ability to ask the right questions. It is an essential tool in continuous improvement.

Chapter 6
IMAGINEERING

The *most powerful* tool of the new management system is imagineering. It enables employees at all levels to visualize how things would be if all problems were eliminated. Since the goal of The Right Way to Manage is to identify and eliminate waste and missed opportunities, imagineering plays a central role in implementing the new management system.

FAMOUS IMAGINEERS

Techniques similar to imagineering have been used by many famous and successful people in all walks of life—businesspeople, athletes, musicians, lawyers, physicians.[12]

Dr. Charles Mayo rehearsed every step of surgery before entering the operating room. He'd mentally call for the scalpel, "feel" the slap in his gloved hand, make the incision, and continue through the entire operation in his mind's eye.

Clarence Darrow often ran through his day in court *before* a trial began. He'd rehearse his arguments, weigh their effect on the jury, anticipate his opponent's strategy, and decide how he would counter it.

Jack Nicklaus sometimes spends up to a minute addressing the ball before taking a shot. Asked why, he responds: "I'm visualizing the perfect shot." When he has a picture clearly in mind—of how the club will feel, whether the ball will start off low or high, where it will land, if it will bounce forward or backward—then and only then is he ready to hit what is nearly always a winning stroke.

The list is virtually endless. Basketball's defensive genius Bill Russell, boxers Gene Tunney and Archie Moore, figure skater Carol Heis, concert pianist Arthur Schnabel, actor Bob Cummings—they all used mental practice to prepare for a major event, overcome bad habits, and establish mental pictures of the right way to do things. They didn't call their practice imagineering, but their mental rehearsal of facts and possibilities was very similar to what we're advocating in this chapter.

WHAT IS IMAGINEERING?

Imagineering is a form of mental practice that requires going over things in advance, visualizing how things would be if everything went just right. Those using imagineering successfully in sports, public speaking, or any endeavor have a clear mental image of the way things should be so that they are prepared for all eventualities.

Imagineering is not a retreat into fantasy but a useful, practical approach to real situations. It is very much grounded in reality and should be backed by enthusiasm and a determination to make the mental image become reality. Although the examples I listed are from the non-business world, the same process can be applied successfully to all facets of business.

The term imagineering was coined at Alcoa Corporation to mean "letting your imagination soar and then engineering it back to earth." My definition of imagineering is visualizing how things would be if everything were perfect. No problems, no errors, no troubles of any kind. Through creating such a mental image, and with a clear and detailed knowledge of the way things are now, you can begin to see why things aren't the way they should be. Once you have identified that "why," you'll find it easy to start working to correct the problem.

Competitors in many fields today use "benchmarking"—comparing their performance to that of a competitor or other organization doing a similar activity. Making such comparisons can be useful, but it can also be limiting, as everyone waits for the other person or the other company to break the current standard. Consider the story of the four-minute mile. When Roger Bannister ran the four-minute mile in Oxford, England, he broke a barrier, set a new standard of performance. Ten years later, eight people did it in one race! If Bannister had been busy benchmarking, he might have been satisfied with just winning races rather than breaking records. So we strive for the ultimate benchmark—UTOPIA.

While the examples above described the mental practice of some very talented people, no special talent is needed to imagine a trouble-free "perfect" process. Ordinary people who are involved in a process can see what it would be like with no problems and then compare that image to the present situation. This is not a brainstorming system which focuses on a particular, narrow problem and which requires especially creative people to achieve results. It is based on facts and allows anyone, armed with those facts, to see how to improve a process.

IMAGINEERING IN ACTION

One company was having a serious problem delivering orders from stock quickly. The time from order receipt to shipment ranged from 2 to 9 days, averaging $3^1/_2$ days. Often some of the wrong merchandise was shipped, causing rework and angering the customer. A major competitor was guaranteeing shipment by the second day after order receipt and was meeting that guarantee over 90% of the time. The company was beginning to work on continuous improvement and decided to try imagineering the problem.

The team assembled for the project imagined the perfect process starting when the orders came in by mail, phone, or fax. All orders from creditworthy customers would go instantaneously to the warehouse, where packers would immediately know the weight of each item and the total order weight and the number and size of boxes needed for the order. In filling the order they would not have to retrace their steps, because the sequence of items on the order form would be the same as in the warehouse. They would make no mistakes in filling the order. The form would indicate the method of shipment, and both the proper number of labels for the boxes and the shipping documents would be ready. Orders received in the morning mail, or by phone or fax before noon, would be shipped the same day. Orders received in the afternoon would be ready for shipment the following morning.

This description of the imagineered perfect process resulted from a series of meetings which included all the people working in the present system: warehouse workers, shippers, foremen, shipping and order entry clerks, credit analysts, data processing specialists, and two managers. These people were not particularly creative, but they did know the process and where the problems occurred. By thinking and working in the new way, they all became creative. They were asked to imagine a system without any problems or errors. They were told to be completely open and not to worry about freely discussing the problems of the present process. The purpose of the session was not to lay blame for past mistakes but to visualize how things could be in the future if everything were perfect. There were no sacred cows protected from discussion. The group became very excited about this new approach and after two meetings came up with a description of the perfect process.

Members then gathered facts about the waiting times in the current process and compared the real to the perfect. They made flow charts of the two processes. They made run charts and Pareto charts of the delay times so they could see when delays occurred and for what reasons. The group assigned members to devise ways of coming as close as possible to the imagineered process for each part of the operation. Within six months, they had established a process that met their goals.

Regular customers now enter their orders directly on a computer or mail them in specially identified envelopes that go straight to the order entry function, which also handles phone and fax orders. Workers enter orders into the computer and verify them. Using established criteria, the computer can approve credit 95% of the time. The computer places the items in the same sequence as the warehouse layout and computes the weight of each package and the number and sizes of boxes needed. The computer also directs a terminal in the warehouse to print out the packing list, shipping labels, and other shipping documents. The packers take the lists, gather the proper number of boxes on a cart and walk through the warehouse, filling the boxes. The warehouse has been reorganized to make their jobs easier, and new box sizes reduce carton costs and give customers less packaging material to dispose of. As packers fill their orders, they use a bar code scanner to gather information about which items they actually pick. When they return to the loading dock, they enter the information from the bar code scanner into the computer, which compares the items selected with the original order, identifying any discrepancies before the boxes are sealed.

The group's hard work has resulted in a process that is very close to the imagineered solution. The company follows the imagineered schedule 96% of the time, while order processing costs have fallen 30%. Most importantly, customer satisfaction went up so much that a downward trend in sales turned around, and during the first year of the imagineered solution, sales increased by 17%.

This example demonstrates a number of fundamental elements of the imagineering process.

1. The company enlisted the help of the people who were operating the process since they were the ones who knew of all the problems.

2. It fostered an open discussion with amnesty given for past mistakes. It eliminated the fear of criticism for things done wrong in the past. The lack of fear meant that people were ready and willing to speak up at all times, allowing for open discussion across departmental and divisional lines.

3. Group members gathered facts about the process, and this data enabled them to locate the waste and determine what to work on.

4. They charted the data to help them understand and communicate it.

5. All the people in the process were involved on a continuing basis. Their imagineering produced not a "program" that would run for a few months and then be forgotten but a new way for everyone to think about the process every day.

6. This approach directed the natural creativity of workers to find the waste and opportunities in all the activities of the organization. I call this "focused creativity." Creativity is focused through the lens of facts and data. The old system of attacking problems often was long on opinions but short on facts.

7. It helped everyone's performance by providing perfection as a target. As Henry David Thoreau said, "People seldom hit what they do not aim at."

THE BENEFITS OF IMAGINEERING

Imagineering pays numerous benefits. The new system leads to better use of all resources, especially brainpower. Imagineering combines the old system's emphasis on mental abilities with the new system's emphasis on working together on the facts and using the information contained in variation.

Imagineering fosters teamwork. Because one person's ideas feed on another's, people learn to work together to solve problems. Most importantly, imagineering helps people visualize any work activity, operation, situation—even a whole business or department—that does nothing but value-added work, with no waste or wasted opportunities of any kind. This image helps to instill the will and belief that something ought to and can be done to fix problems. It helps people know what to work on to get results.

KEYS TO SUCCESSFUL IMAGINEERING

For successful imagineering

1. Establish and maintain an open atmosphere.
2. Ground imagination in reality and base it on facts and data. Only people who know the current process well can successfully imagineer a perfect process.
3. Make "pervasive imagineering" a constant, ongoing approach to problems.

In the old system, discussions at meetings are often dominated by a few people—those with strong personalities or selling skills, or those in a position of power. In the new system, all people contribute to the discussion. Since the discussion centers on facts, no one wins any points by sheer force of personality. Instead, the strength of *ideas* wins points. People should provide empirical evidence to support their points rather than rely on judgments arising from emotions, pet peeves, or pet projects.

To achieve this kind of open atmosphere where no one is afraid to contribute, leaders need to drive out fear. People need to believe they will have a chance to put their ideas on the table without worrying about criticism or retribution. Everyone deserves amnesty in the new management system. The issue is not who caused the problem but how to correct it and avoid the same or similar problems in the future. The focus is on fixing the system, the process, and not on assigning blame. This kind of open atmosphere is helpful not just at meetings, but throughout the organization. Workers and managers—working alone or in groups—should feel free to look into troubles, opportunities, and waste, and to imagineer how things could or should be.

People should develop the habits of basing their imagineering on facts and turning the facts into data whenever possible. Facts help people understand the process, and this understanding in turn promotes more effective imagineering. Saying we had a lot of waste last week may be a fact, but "a lot" is very subjective. Showing a breakdown of the waste by cause on a Pareto chart eliminates ambiguity.

Leaders in the new management system provide people with ample opportunity to prepare themselves, to get the facts to present. In the old system, managers might appear in the morning and announce a 1:30 p.m. meeting. That may be fine in some instances, but if the people have a chance to collect data on specific issues to be discussed, the meeting will be more productive.

Because imagineering sessions in the new system deal heavily with facts, they encourage effective, focused analysis. People spend their time looking at the facts and talking about where the facts lead them—rather than rambling through a litany of possibilities and opinions. For example, a group imagineering about an excessive inventory problem would present specific facts and data—what makes up the inventory, how long do particular items sit, what processes would suffer without the inventory. These facts in turn could lead to specific ideas for reducing the inventory.

Successful imagineering leads to action. The outcome of an imagineering session should be an action list showing who will do what, what additional data must be gathered, and what has the highest priority. Of course, imagineering should be going on in the minds of participants all the time between meetings as well as during meetings.

People can be trained and conditioned to imagineer on their own as well as part of a group. Spontaneity is crucial to the success of imagineering. While they are working, people should be thinking of ways their work could be improved. Everyone should encourage the use of imagineering at every stage of a project.

Eventually this technique can pervade everyone's thinking throughout the organization. Success in the new management system depends upon management making imagineering part of the corporate atmosphere. With its advertising slogan that it "never stops asking what if . . . ," Hewlett Packard declares to the world that continuous imagineering governs the way its employees think and work everyday. The ad implies that with imagineering as a process, innovation is sure to be a product.

It's evident why imagineering represents such a powerful tool in the new management system. Waste is not only difficult for managers to see, it's also difficult for many managers to *confront.* If we define waste as the difference between the way things are now and the way they would be if everything were perfect, then it is virtually impossible to eliminate waste without imagineering. Imagineering opens the doors—and people's minds—to waste, problems, and opportunities, providing the key to improved productivity and quality. It releases the power of the people to eliminate waste in the processes which they operate.

Chapter 7
PROJECTS

To translate insights and ideas from imagineering into concrete results requires a lot of work. This work is best accomplished through a series of projects, each focusing on a specific goal. Ideas for these projects come both from the top down and from the bottom up. They come from imagineering and fishboning sessions. They come from asking the experts, from surveys of customers, suppliers, and employees, from listing problems and opportunities on Pareto charts. They come from the open atmosphere in the new system, in which people are looking for continuous improvement without worrying about blaming anyone for the mistakes of the past. Most especially, they come from identifying and quantifying waste, the subject of the next chapter.

Project teams, either formal or informal, undertake most of this work. Management appoints formal teams and designates the teams' leaders. Informal, ad hoc teams may arise to solve specific problems.

I classify the project work into three major types: 1) major management-directed projects and subprojects, 2) personal projects of senior management, and 3) other projects.

MAJOR MANAGEMENT-DIRECTED PROJECTS

Unfortunately, most organizations don't have the resources to support all good project ideas. To maintain enthusiasm, the organization needs to give people some amount of freedom to work on the things they consider most fruitful or interesting. Yet management must insure that the most important and urgent projects receive the required resources. Management therefore needs to designate important projects in some special way so that everyone understands that these projects deserve top priority. We call these "major management-directed projects." Such projects are directed by senior management and are so large that they require the assistance of most people in the organization and bring important change to the unit. Their importance is clear to everyone—suppliers, customers and all employees. They can include large and significant quality improvements, major cost reductions, important new product or technology developments, substantial changes in the time required to do important things, and major alterations in marketing or distribution. These projects change the business significantly.

If these projects are to succeed, they need more than just a special name; they must literally be directed by management. Some member(s) of senior management should be deeply involved in each project and feel personally responsible for its success. The manager need not be the leader of the project but must at least act as an "enabler," insuring that resources are available and roadblocks are eliminated. The unit involved should have the infrastructure for continuous improvement substantially in place before attempting this type of program.

Obviously such major projects generate a substantial number of sub-projects. The overall leader should coordinate the sub-project teams to insure they are supporting the main project. He or she must establish a system of measurement and reporting that focuses on the key operating variables.

PERSONAL PROJECTS

Individual managers also should have their own personal project(s). By assuming responsibility and doing the work themselves, rather than delegating, managers signal the whole organization that this is the new way of working for everyone.

About four or five years ago, I suggested to the executive vice president of a large Fortune 100 company that he could help people get started to work the new way by leading a project himself. I encouraged him to pick one that people would see and believe in, one in which he could use the simple tools, make some quick and visible gains, and be personally very visible. I wanted his people to see that by working the new way they could make important progress at all levels and do it quickly and well.

The vice president—I'll call him Joe—selected reducing the use of all forms of energy. He chose this project because he could involve a wide range of people—office people, administrative people, secretaries, maintenance workers, plant engineers, managers everywhere. The project's focus included air conditioning at corporate headquarters in the South; lighting; people's use of gas and electricity in the plant; energy efficiency of the machinery; everything.

During the next two years, 300 to 400 people under Joe actively worked on projects and sub-projects about energy. Each one used the simple charts, collected data about energy use in a particular area and broke it down into its smallest details, found ways to do the work better while using less energy, put in devices to shut off the lights automatically in offices when no one was there, set the optimum temperature for the air conditioning, questioned how well the air conditioning delivered what it should . . . all over the company.

Someone—a secretary, an engineer, a manager—was in charge of each project. He or she would coordinate the data collection, keep track of progress, find out what could be done, and lead the necessary changes.

When Joe's organization began the project it performed a survey of all energy uses and found that it was spending $6.2 million annually on energy. Over the next year and a half to two years it actually reduced that figure by over $3 million annually.

Everyone recognized the tremendous accomplishment of saving $3 million, nearly half the energy cost. But far more important for the long run good of the company were that (1) hundreds of people understood that top management was seriously involved, ready and willing to lead the projects; (2) all learned how to use the simple charts, to use data collection techniques, to follow the data until they found the waste, and to set up systems so that they could maintain the gains by keeping charts; (3) everyone worked on teams, communicating well to keep others informed throughout the area. All these people learned that they could make a difference. They could do something significant and learn to do more. They were now ready to work on other projects throughout the organization. This was an outstanding example of what a leader can do to help a lot of people in the organization start successfully working in the new way.

OTHER PROJECTS

Besides managers' personal projects and the major management-directed projects and sub-projects, numerous other projects should arise throughout the organization. Individuals or small groups should feel free to develop projects on their own, but they should recognize that they may not always get the resources they need when they need them. It is up to management to insure that the people who want to do projects have been properly trained in gathering data and in the tools of variation. Managers in general should make it a point to know what is going on in the projects in their area and to give as much encouragement and help as possible. For the new system to succeed it is important that project work be done everywhere, by the people closest to the work, whenever they see a need.

These projects generally require a minimum of formalized reporting, since any hint of bureaucratic control tends to stifle enthusiasm. The organization should have some sort of system to give recognition to the participants—team members—in successful projects. The recognition given to successful projects feeds the enthusiasm of people to work on continuous improvement.

The next two chapters explore project work in more detail. Chapter 8 focuses specifically on developing projects to eliminate waste throughout the organization. Chapter 9 presents a more general overview, a method of process improvement developed by Conway Quality that leads people through the improvement process with a minimum of wasted effort.

Chapter 8

IDENTIFYING, QUANTIFYING, AND ELIMINATING WASTE

Because the core activity of The Right Way to Manage is to identify, quantify, and eliminate waste through process improvement, I'm devoting an entire chapter to waste. Historically, most businesspeople have thought of waste as, in one dictionary's definition, "damaged, defective or superfluous material produced by a manufacturing process." This is a very narrow definition, covering only a tiny portion of total waste. Another definition in the same dictionary comes much closer to my own: "to waste is to allow to be used inefficiently." My definition is broader still: "Waste is the difference between the way things are now and the way they would be if everything were perfect." Since things are never perfect, the hunt for waste must be continuous and ongoing, **forever.**

By my definition, the amount of waste is huge. A few years ago, a Texas newspaper ran the headline, "Corporate Waste Estimated at $862 Billion." At the time that figure represented 24.6% of the Gross National Product of the United States. Today corporate waste exceeds one trillion dollars! My own experience has led me to conclude that most U.S. companies waste an amount equal to 20–50% of their sales with a median of about 40%. People have a great deal of difficulty accepting these numbers until they figure them out for themselves.

One person who had a hard time believing my figures about waste was Paul Pankratz, Vice President of Operations for Dow Chemical USA, whom I first met in 1984. Paul came to a talk I gave at his company and decided to work with Conway Quality. But he assured me that my waste estimates didn't apply to his company, which he considered one of the best managed companies in the world. He expected that Dow might have just 7% or 8% waste, but because he ran a $4 billion operation, and 7% or 8% would amount to around $300 million a year, he thought working harder to reduce waste would still be worthwhile. So he hired Conway Quality to help him get rid of that 7 or 8%.

I didn't argue with him at the time but instead told him a story which I will come to shortly. Dow went after waste, especially at its plant in Texas. In their first full year of operations under The Right Way to Manage, the company's Texas operations saved $44 million, a figure announced at the first NASA Annual Quality Conference in 1986. In the following year, it saved an additional $100 million in its operations, much of it in Texas. After only two years its total savings amounted to some 5–7% of net sales.

About two years later I had occasion to ride with my old friend Paul who said, "After working the 'new way' two years, and seeing the progress we made, you know what I think the waste is now, Bill? I think the waste now is somewhere around 30% or more. As soon as we learned about the work, we saw the waste was everyplace."

I said, "Paul, remember that story that I told you two years ago? 'Waste is like an onion; peel back the first layer and you look and say, oh no, look at all that waste. Peel back the next layer, it looks even worse. Peel back the next, it looks worse still. The closer we get to the real work, be it the work of machines, chemical processes, computers, people, energy, or any kind of work, the worse it looks.'"

How do you go about identifying this waste? Obviously since waste is the difference between how things are and how they should be, imagineering is a particularly helpful tool. Because there is waste in all processes, you look everywhere for facts with which to begin your imagineering. You look for waste in production, administration, distribution, sales, marketing, R&D, service, everywhere.

To help think about the waste, classify it into four types—waste of capital, waste of material, waste from lost opportunities or sales, and waste of people's time, energy and talent. As you gather facts, quantify the waste as you classify it, to insure that you are dealing with facts and to help you understand what is happening. Because every organization has finite resources it's impossible to tackle all forms of waste at once. Usually, the areas with the largest amounts of waste require the highest priority.

Every time you identify substantial waste, attach a dollar amount to it. In some cases, the dollar amount can be a very precise calculation, while in others it is a rough estimate. The cost of waste in a box of scrap can be calculated fairly accurately. But the waste associated with product delays, excess inventory, and unwise use of energy may require rough estimates.

However rough, these calculations are essential, because they point to the most important sources of waste. As Lord Kelvin said, "When you can measure what you are speaking about, and express it in numbers, you know something about it; but when you cannot (or do not) measure it, when you cannot express it in numbers, your knowledge is of a meager and unsatisfactory kind."[13]

As we have emphasized throughout this book, you should use the tools of variation to help understand and quantify the waste. In the following sections, we will discuss how to use those tools to identify, quantify, and eliminate the four major types of waste.

WASTE OF MATERIAL

Wasted material is the most obvious form of waste. Most people equate this kind of waste with scrap, particularly in a manufacturing process. Product sold as seconds is a slightly less obvious form of waste. Wasted energy can't be seen, so it is often ignored.

An even more subtle form of waste is caused by specifying materials that don't meet customers' requirements. Specifications that are too strict and therefore needlessly expensive can be as wasteful as specifications that are too lax and lead to rejection of the product. A value analysis can reduce this form of waste by making clear what characteristics are really needed in purchased and produced materials. Too many organizations design and build products with insufficient understanding of customers' needs and wants. Constant communication with customers, potential customers, and suppliers is critical. Any work process that is not designed to meet customers' needs contains waste.

Variation in the material used is also a great source of waste. Suppliers should use the tools of variation, particularly statistical process control, to minimize product variation and to insure their process can operate within your specification limits. And of course your own processes must be under statistical control so that you can address problems as they arise, not after a lot of bad product has been made.

Even finding the relatively obvious material waste is not easy. It requires looking at your processes differently and in a lot more detail than you have done before. People do not see that the waste is all around them until they study what they are doing in detail, particularly as it relates to customer needs. I find that you need to get people thinking in a different way about waste to discover it.

A good example of this different kind of thinking occurred in mid-1981, when Jim Copley was General Manager of a $60 million business at Nashua Corporation. Jim's division was making and selling toners, developers, and other supplies for copy machines.

Along with his senior managers and supervisors, Jim had been to Dr. Deming's four-day seminar. They had all undergone the training on simple charts, and Jim reported every month about some quality project on which he and his staff were focusing. They were doing some good work in quality.

However, by this time a few of us really knew what to do and how to do it, and I didn't think Jim's division had done enough to combat its waste. So I telephoned Jim and told him I'd like to help him really get started in working the new way. When I explained that the new way meant working on quality, he insisted that he already was and cited examples. I told him I was glad he had made a start.

When we got together, Jim was feeling good because he expected the business to make $5 million pre-tax for the year on $60 million of sales. I asked, "Where does the other $55 million go? We need some of the $55 million. In addition to selling more stuff, we want to have more of that 55."

Jim got out the budget book and started to tell me things from the budget.

I said, "No, Jim, we need to know a lot more about it than what's in the budget book. That is just the financial numbers— results. I want to know in detail where all the waste, the opportunities for improvements, exist in this business."

Over the weekend Jim did some work with the Controller and some of his other people to come up with their waste. By our next meeting on Monday, they had found approximately $7–8 million of waste. They found waste in material, inventory, everywhere.

I praised Jim for a good start, and we called the other managers in from Manufacturing, Sales and Marketing, Administration, and Customer Service. I suggested that to continue to search we should dig deeply into one area and get an idea of what kind of waste really exists. I asked Jim what area he wanted to search—Sales, R&D, Administration—and Jim picked Manufacturing. Manufacturing tends to be a bit easier than other areas. People are used to keeping numbers in manufacturing so they can identify waste more easily. They can see the physical waste in their plants and are used to working on waste reduction.

The cost of goods sold was approximately $30 million, or 50% of sales. Jim and I and others set out with Mike, the Manager of Manufacturing, to determine where all the waste was in that $30 million. We went over his area in detail as we asked many specific questions about work and waste. What about the time of the people? Have we got the right equipment? Does it run perfectly? What about the material waste? I noted that you could see black toner sticking to the walls. I asked other questions about safety and the environment. We also discussed over- and under-specifications. Obviously we had some waste everywhere.

When we got together again about six weeks later they had found, in the manufacturing area alone, about $10 million of waste. They saw it everywhere. To find waste of material, they started with the raw materials and compared them to what ended up in the packages for the customer. They had been overspecifying in some areas, and they had wasted people's time. They now saw that they could run the business well with about half of their current 100 employees.

I thanked them for a job very well done and said, "Jim and Mike, we're still not deep enough into the work. We're not deep enough down to know what we need to know. We're a lot better off than we were when there were just Jim, Mike, and the Controller looking at it. Now with Mike and all of his people we are really starting to dig in and seeing a lot more. Mike, let's pick one of the areas in your business, people's time, raw materials, packaging materials, inventory, what would you like to work on?"

Mike said, "Packaging material." In his business, they had a lot of it. The products were used in offices so there was special packaging—lots of small boxes, bottles of toner, caps and labels. We were using $2.3 million of packaging materials a year.

In addition to looking at the actual materials, Mike and his team examined the workers' time, utilization of the equipment, maintenance—all the costs associated with doing the packaging, $1.5 million of costs. They got a large conference room and laid out all the packaging materials. They knew the quantity used and the price they were paying. They had detailed information. The marketing, manufacturing, selling, and packaging people looked at every single package. Is this the right size bottle? Does it need to be upgraded in order to sell more? Are there cost savings? Where? What? How? Should it stay the same? They looked at everything in the most excruciating detail. The study group included people at all levels from all areas—marketing, engineering, manufacturing, finance, sales, customer service, etc.

They studied all the packaging materials in depth. They clearly identified $900,000 that they could save in material costs by eliminating some things, changing others, and doing others less expensively. In less than a year they reduced their packaging material costs by over $700,000. They also saw that out of $1.5 million they were spending on the packaging machines, they could save around $700,000. They used bottles of different sizes and shapes; they came up with ideas to help sales sell more to customers; they found opportunities for cost reduction; they suggested ways to make products more appealing and useful to customers. They used larger cartons and smaller cartons—whatever was optimum. Everyone worked together.

At the end of five months we had identified $1.6 million in waste out of the total $3.8 million packaging costs, and we were continuing to see opportunities for improvement. Everyone now saw the value of digging deeply into the work to find the waste. At the beginning, they thought they had 12% waste. At that point, they had done what was for most people a good job at finding the waste. As soon as they looked closely at the manufacturing area, what happened to that waste? It was over 30%! As they looked even more deeply into the packaging materials themselves, they found a total of 42% waste.

We were excited. Everyone was learning together. Jim Copley and his team of managers, supervisors, and other employees had now learned that detailed studies of work and of work processes led to the waste. They went on to do similar projects in such areas as marketing, sales, R&D, administration, and finance, finding huge waste everywhere and attacking it. Jim is now helping companies worldwide as a Senior Associate of Conway Quality, Inc.

WASTE OF CAPITAL

Three major uses of capital that organizations should examine for waste are receivables, inventory, and plant and equipment. The waste is not often obvious in these areas, but one way of discovering it is to see if the capital is idle or if it is active. Idle capital is waste. Receivables not collected on time is idle capital. So is inventory sitting in warehouses or on the shop floor and equipment not being used full time. Shareholders invest capital to make money. If that capital is not being fully utilized to make a product or service, it is being wasted.

The conventional wisdom about receivables is that they don't get paid on time because customers drag their feet. In my experience of working with numerous companies, about 70% of late payments are due to supplier errors. To find out why your receivables are not being paid, call the customers, not to dun them, but as part of a survey to find out why customers don't pay on time. Explain that the purpose of the call is to find the problems and fix them. Time after time I have discovered that the bill was sent to the wrong place, the product or service was not what was expected, or the paperwork was full of errors or did not match the customer's needs. In every case problems within the supplier's control caused more than 50% of the late payments.

One of Nashua Corporation's very large, financially strong customers paid regularly in 60 days, when the terms were 2% discount for payment in ten days, net due in thirty days. When asked, the customer said that they wanted to take the discount, but by the time the invoice arrived and was processed through the various departments, it was more than ten days from the invoice date, so they put it in the pile to be paid in sixty days. By making special arrangements to have invoice copies sent directly to all the departments involved, they were able to take the discount and Nashua received payment in ten days instead of sixty. The customer saved money and with interest rates well over 10% and cash in short supply, Nashua was happy with the arrangement. Simply understanding the customer's needs benefited everyone.

Idle inventory is wasteful in many ways. It uses capital and storage space, it must be counted periodically, and it is subject to damage and obsolescence. Furthermore, it hides errors. Inventory of work-in-process may contain a lot of bad material which won't be discovered until it is made into a finished product. The more inventory you have, the more likely you are to have waste hidden in it. The techniques of "just-in-time" inventory, as originally developed by Toyota, work very well to eliminate idle inventory. To learn about the techniques, read *Workplace Management* by Taiichi Ohno[14] and *Zero Inventories* by R. W. Hall.[15]

Examine equipment for waste from excess capacity, bottleneck operations, unnecessary backup, and over-specification. Extra capacity, like excess inventory, often makes people feel comfortable, but by using temporary extra shifts, subcontracting, and proper planning, companies can and should closely match capacity to needs. Too little capacity, which is also a waste, often results from bottleneck operations. Before adding equipment to eliminate a bottleneck, make sure that the present bottleneck equipment is operating full time at rated speed and with the proper maintenance and that employees are properly trained to operate it.

The comfort factor also leads to unnecessary backup and over-specification. The sin of not being able to produce or deliver is much more visible than the sin of spending too much for equipment, so industry is loaded with extra equipment and extra capability "just in case." People must understand that they are responsible for maintaining a proper balance between equipment capability and need. Buying equipment just to be "safe" should not be acceptable except in very unusual circumstances.

WASTE OF PEOPLE'S TIME, ENERGY, AND TALENT

Underlying all the waste in an organization is the waste of human resources. Most organizations have available the talent and brainpower to eliminate most waste and bring about amazing improvements in their ability to meet customers' needs. Almost all people want to be doing full-time, meaningful, value-added work. They would like to feel that their organization operates as a team, with each person a valuable member of that team. They would like to make important contributions to the success of the organization. It is a grievous failure of Western management that too few working people are given the opportunity to participate in this manner. Their brainpower and energy is wasted, as is their time.

"Restructuring" in U.S. companies strikes fear into the hearts of employees because it has come to be a euphemism for "head-chopping," a reduction in force. Such programs often save large amounts of money by squeezing out waste that has never been identified. But they have a chilling effect on the employees, who recognize that they are being made to suffer for what is, in most cases, the failure of management. Often restructuring is analogous to slash-and-burn agriculture, which is so poorly planned that a given area will be productive for only a few years, after which another section of forest must be destroyed to make more temporary farmland. If an organization doesn't change its work processes and hiring practices, if it keeps working in the old way, the waste will come back and it will need to restructure again. A major American corporation slashed its white collar payroll in half in 1983, yet by the end of 1989 it had more white collar workers than before the restructuring. And so it slashed again.

A need for restructuring implies that management allowed waste to accumulate and failed to create a system that gave people full-time, meaningful work that created value for customers. That type of management can never provide employment security. In a market economy, only those firms that meet customers' needs with the least amount of waste can provide employment security. To offer such security, management must work in the new mode of continuous improvement—eliminating waste and preventing its return.

WASTE FROM LOST SALES OR OPPORTUNITIES

Many organizations acutely suffer from lost sales that could have been made and lost opportunities that could have given the organization a boost. Organizations lose sales for any number of reasons, including product problems, pricing problems, poor customer support, bad market positioning, insufficient or misdirected sales effort, and lack of communication with the customer. Non-commercial organizations that do not have sales per se also suffer from lost opportunities. Colleges do not attract the applicants they want because they miss opportunities to learn what students are looking for and tailor their programs to meet those needs. Government agencies may waste taxpayers' money because they neglect to search for ways to serve their clients better, whether those clients are taxpayers or other government agencies.

I have found three techniques invaluable for boosting sales and uncovering opportunities: lost order reports, analysis of variation of sales, and analysis of the work done by sales and marketing people.

Create lost order reports. The people who know why a company loses orders are the salespeople and the customers. You will get a good feeling for what's going wrong if you ask your sales people, as part of their regular call reports, to describe business that they lost for any reason and to summarize those reasons. This technique also makes salespeople think about their efforts in a different light and perhaps begin to see how they could do things differently. If you think sales were lost because salespeople didn't call on potential customers, enlist marketing people to survey those potential customers and find out which ones should be called or visited.

Having compiled a list of reasons sales have been lost, you can begin a series of projects to eliminate those reasons. All areas of the company—not just sales and marketing, but also manufacturing, administration, R&D, maintenance, customer service, product development—need to get involved in eliminating the reasons customers don't buy.

Analyze the variation in sales patterns. You can most easily see variation if you compile information about each sales territory's market share in each product line. Try to discover the reasons for that variation, so you can increase your market share. Look for special causes of variation that can be eliminated and for common causes that are a part of your sales and marketing system. Search for variation that may be due to sales training, level of customer support, and type and level of advertising and sales promotion. Again, use the information to identify and set priorities for projects that can narrow the range and increase the level of sales performance in all areas.

Analyze the work done by sales and marketing people. Just how to accomplish this will be covered in the next section describing the waste of people's time. The major value-added activities of salespeople are planning sales contacts, contacting customers and potential customers, providing information about customers' needs and wants, and acquiring product knowledge and selling skills. By asking salespeople to do a work-sampling study on themselves, you will likely find they are spending a relatively small proportion of their time on value-added work. In my experience, the average sales force spends less than 30% of its time in those activities. Salespeople usually spend a lot of time working on product or delivery problems or calling on the wrong customers. Salespeople are generally amazed when they analyze their own use of time and find how little they spend on actual selling. They are happy to cooperate once they see how the analysis can help them improve sales performance.

The data from lost order reports, analysis of variation, and analysis of salespeople's work can lead you to the problems. A well coordinated group of projects to eliminate those problems can bring about immense improvements in sales efficiency.

With an understanding of the tools of The Right Way to Manage and how they can be used to combat waste, you are ready to tackle the methodology presented in the next chapter.

Chapter 9

PROCESS IMPROVEMENT METHODOLOGY

Obviously no "cookbook" guide to continuous improvement could succeed for all projects in all organizations. However, we have developed a methodology for attacking waste that we have found to be extremely useful in getting people to work on projects in a logical sequence. It ties together all the tools we have introduced in the past six chapters.

This methodology provides a common language and a set of tools for project teams to use. Teams can follow defined procedures to identify and eliminate the waste. These procedures answer the questions, "What do we do now?" and "What do we do next?" Getting people to use the language, tools, and procedures every day, in all their continuous improvement activities, is part of the cultural change required in switching to the new way of management.

The twelve steps in the methodology are shown on the following page (Figure 9-1). All steps may not be required for all projects. In fact you will discover, as you identify the waste, that some things are relatively easy to fix, and if it is obvious how to fix the fundamental problem, you should do so as soon as possible without mechanically going through the other steps in the methodology. You will be amazed how many things fit in this category. Waste has a way of escaping your notice because it is part of the routine. Until you look for it in a methodical way, it seldom comes to light. Remember, however, that if you do not change the work process, the trouble will come back.

Figure 9-1

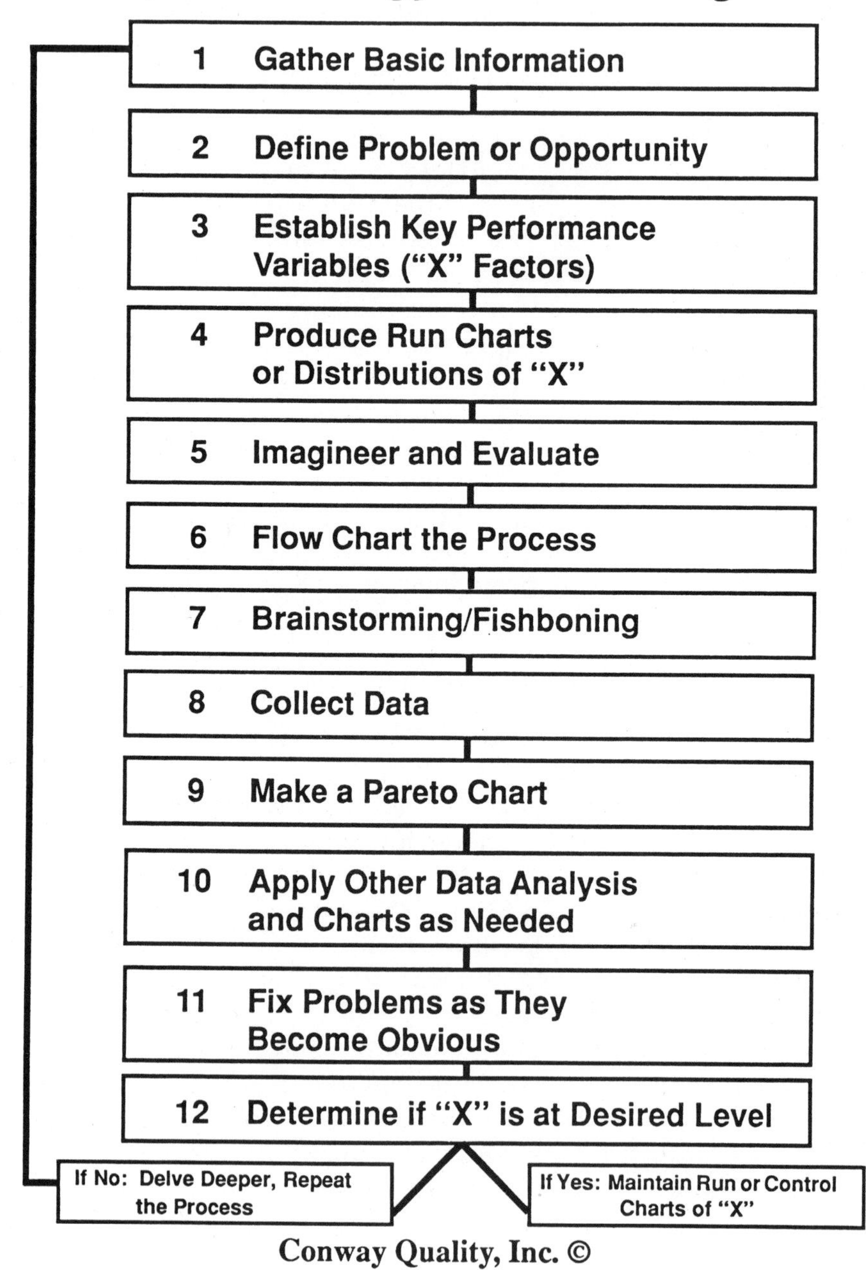

Conway Quality, Inc. ©

1. *GATHER BASIC INFORMATION*

Gather information in the ways discussed in Chapters 3, 4, 5, and 8: 1) collect data, 2) survey customers, suppliers, and the people doing the work, and 3) study the work and the work processes.

2. *DEFINE THE PROBLEM OR OPPORTUNITY*

The more specific you can be, the better. Describe the waste to be attacked and assign it a monetary value, even if it is only an estimate, to help set priorities and to measure progress. Remember, waste includes wasted opportunities. As part of ensuring that everyone is using a common language to talk about and work on the same thing, Step 2 should include all-important operational definitions of the waste. For example, if you are talking about waste from excess inventory, you need to define what to include in the project: what product lines, what locations, what raw materials. And what about work-in-process, finished goods, goods on consignment, goods in transit, operating supplies, maintenance supplies, office supplies? Depending on the scope of the project it may be appropriate to include one, several, or all of the items mentioned. But everyone involved must understand what you are working on. Hence the need for an operational definition.

3. *ESTABLISH KEY PERFORMANCE VARIABLES*

This is a critical and sometimes difficult step, although it can be simple for straightforward operations. For preparation of invoices, for instance, the number of invoices prepared per hour and the number of errors per hundred invoices are obvious measures of performance. Less obvious factors may affect the process in more complex operations. For example, the ambient temperature and humidity may have a profound effect on some manufacturing processes. But you won't know for sure until you perform some experiments. Try to isolate the key factors that correlate with process performance. If you are specific in Step 2, you will have an easier time establishing what to measure.

These factors need to be measurable, preferably on a real-time basis, so that corrective action can be taken immediately if one of the factors goes out of control. Devising how to measure the critical factors may require some ingenuity. For example, invoice errors may not become known until much later, if at all, unless you can devise some type of checking mechanism. You also may need to break a performance factor down into sub-factors. Invoice errors may need to be classified as pricing errors, data entry errors, and product identification errors, since they may have different causes.

Performance factors are commonly measured in time, money, quality, volume and per cent. Identifying these factors is worth a lot of effort, because when you are measuring the right things you are more than halfway toward finding the causes of the waste.

4. *PRODUCE RUN CHARTS OR DISTRIBUTIONS OF KEY PERFORMANCE VARIABLES THAT AFFECT THE PROCESS*

Often the obvious place to begin is with the run chart of how the factor varies with time. This chart provides an initial feel for how the factor is varying and may give clues as to why it varies, particularly if the chart is kept on a real-time basis. If historical data is available, it, too, can be charted, providing a head start in understanding how the process varies.

In some cases it may be immediately obvious that a histogram or correlation chart is needed to clarify how a factor operates. These charts also help the project team talk about the problems.

5. *IMAGINEER AND EVALUATE*

At this point you should have enough facts to imagineer the process. Imagine how the whole system could be and concentrate on the key performance factors you have identified. What would each factor look like if everything were perfect? What would its level be on the run chart, and what would the range of variation be? Evaluate how far you are from perfection and what it might take to make each factor perfect.

6. *ANALYZE THE PROCESS WITH A FLOW CHART*

A flow chart will help you to understand fully the process and how the key performance factors affect it. After creating a flow chart for the present process, imagine what an equivalent chart of the perfect process would look like. This comparison will give you additional ideas for process improvement. There might be two, five, ten processes to chart—complex charts or simple ones.

7. *BRAINSTORM/FISHBONE*

Earlier steps will have amassed sufficient data and ideas about the process to warrant a brainstorming/fishboning session. The cause-and-effect diagram (fishbone) helps you understand the process and focus on problems and opportunities for process improvement. Include in this discussion a number of people working in the process plus some outsiders who may have specialized knowledge. Participation helps the people working in the process to buy in to the need for change.

8. *COLLECT MORE DATA*

You will undoubtedly require additional data to define further the necessary changes identified in the previous step. You may have to get information from the process's customers and suppliers as well as its operators.

9. *USE A PARETO CHART*

At this point you need to set priorities for action. Use a Pareto chart to rank the contributors to waste and to rank the problems and opportunities. Identify the most important projects and sub-projects. Assign responsibility for each project. Make sure the project goal is very specific and is measurable. No one ever fixed a generality. The specifics might include a desired level and range of variation for each of the key performance variables, but never put a cap on quality or productivity. Make trial changes in the process, if necessary, to determine what factors have significant effects.

10. *APPLY OTHER DATA ANALYSIS AND CHARTS AS NEEDED*

Additional statistical tools may be appropriate depending on the type and importance of the problem or opportunity and the data available. Stratification of the data, histograms, correlation charts and/or statistical control charts turn the data into different forms of information and reveal different strengths and weaknesses of the process. Statistical control helps sort random events from non-random or common causes from special causes. See Chapter 4.

11. *FIX PROBLEMS AS THEY BECOME OBVIOUS*

As we mentioned at the beginning of the chapter, you should be doing this all along. If you know how to fix a problem and it doesn't require significant resources to fix it, don't wait. Even small successes help build momentum for working in the new way. Make sure, however, that you are fixing the work process, not just alleviating a symptom.

12. *DETERMINE IF THE KEY PERFORMANCE VARIABLES ARE AT THE DESIRED LEVEL*

If a key performance factor is where you want it, use a run chart or statistical control chart to monitor that level continually. Standardize the process so everyone does it the same way. If the performance factor changes significantly from the target level, investigate the cause and take corrective action. The best and easiest time to fix a problem is when it first appears.

If you have been unable to bring a factor to the desired level, you need to go deeper and repeat all or part of the methodology. You may need to start with another fishboning session on this specific variable. Or you may have to go back to the beginning, gather more information and redefine the problem.

THE DEMING CYCLE

A simplified form of the methodology which is easy to remember is the Deming cycle of PDSA—Plan, Do, Study, Act.

Plan—Gather facts and decide what factor(s) need to be improved and what changes might cause an improvement.

Do—Try a change, preferably on a small scale.

Study—Evaluate the results of the change.

Act—Implement the change or plan a new test, based on the facts you have developed.

Repeat the cycle.

SECTION III

HUMAN RELATIONS IN THE NEW SYSTEM

How do you avoid the most important waste of all in an organization—the waste of human talent? Imagine what perfect human relations would be like in an organization. The key is how individuals feel about themselves and the organization.

In the ideal organization, each individual would be an enthusiastic and valued member of a winning team. His or her major objective would be to help the team be the best in every way. Each person would contribute to the mission of the organization to the best of his or her ability in concert with other team members. The leaders of the team would act as coaches and would provide the tools and training necessary to bring each team member up to full potential. They would enable, empower, entrust.

Team members would respect each other and share their knowledge. Team members would have a sense of self worth and pride because a winning organization was recognizing and fully using their talents. They would be happy to come to work in the morning, realizing that they had control over their part of the process, that they were contributing to continuous improvements, and that their peers and leaders valued their thoughts and opinions. They would feel that promoting the interests of the company was also in their self-interest. Feeling like an important member of a winning team would satisfy and motivate them.

Compare this with the traditional management-versus-employees, "them vs. us," attitude in the old system, and it is easy to see why most organizations waste so much human talent.

When people realize it is in their self-interest to work in the new system of continuous improvement, miracles can and do happen. I have cited numerous examples of major improvements being made using the tools of variation and imagineering. Such improvements are possible only if management is trying to foster the attitudes just described. The next four chapters detail how to expedite the massive transformation in human relations that is an important part of The Right Way to Manage.

Chapter 10

CREATING THE NEW CULTURE

In the new system, all human relations, all interactions in all directions are considered important and should be consistent. In the new system individuals cross departmental barriers to cooperate on projects; all employees understand that working as a team toward continuous improvement is in their own self-interest. Everyone shares a "we" attitude that recognizes that workers and their leaders are both components of the same system, that each component needs support from the other to succeed, and that the people doing the work are the real experts on that work.

This means a different management style. It means developing effective alternatives to the traditional pyramid management structure, working horizontally with customers and suppliers and teammates rather than vertically with "superiors" and "subordinates." It means a series of teams being formed in many different areas and levels of the organization. It means managers leading by helping, coaching, enabling, empowering, entrusting, raising the average, rather than directing and judging.

Generating the proper human relations culture is a key element in The Right Way to Manage. Changing the way people relate to their organizations has to be a major part of any new management system, because people's **brainpower** and **dedication** is the most **under-utilized** asset in Western organizations. Organizations foster that dedication by asking employees for their help and advice and by including them in the decision-making process. Once they see that their organizations really want their help and value their opinions and their expertise, people buy in to the decisions made and to the concept of continuous improvement. Even more importantly, they feel "ownership" of their job. They believe it is up to them to be contributing members of the team working toward continuous improvement. They feel a sense of pride in their work. They are happy to have a chance to use their brains to make suggestions, knowing that their ideas will be valued. For people to feel this way, the leaders of the change must be sincere. They must be working in the new way themselves, and they must believe that they need the ideas and suggestions of all the people. People running the processes quickly detect any lip service in their relationships with management.

Proper training, the team feeling, taking full advantage of everyone's brainpower and dedication—these are the things that release people's power and bring about the miracles in the new system.

Organizations need a human relations system that supports this new teamwork attitude. The system serves as a vehicle for communicating and reinforcing desired attitudes, behaviors, and actions. This system consists of the accepted norms for personal interaction and treatment of people, norms that encourage and insure behavior that supports the organization's goals.

The two components of a human relations system are the personal aspects (the way people interact) and the organizational aspects (the program of rewards, advancement, recognition, and education).

Creating a human relations system that supports continuous improvement requires increasing the quality of interaction at all levels in the organization. The managers must set an example with positive personal behavior that others will emulate. Leadership can start at any level—the higher the better—and it spreads through all levels in the organization. People at every level need to be ready, willing, and able to act as leaders of the change. The organization must remove barriers to continuous improvement from its policies and procedures. And it must adjust the rewards, advancement, recognition, and education programs.

Even when people fully understand all of the technical aspects of working in the new way, the organization will not make huge, sustainable gains unless significant improvements in the human relations system are also taking place. The human relations system tells people how they will get ahead, how they will gain recognition, and how they should treat others and expect to be treated. It confirms or denies to people at every level whether the organization is serious about adopting the new management system.

In the new management system, human relations activities foster loyalty and attachment to the organization. Everyone knows of someone who has turned down an attractive job offer because the individual was happy at his or her current job. Organizations inspire such feelings by creating a sense of teamwork and providing a quality of life that the individual feels will not be duplicated elsewhere.

BARRIERS

When I am at public seminars or visiting companies, I often ask what barriers keep people from working in the new way. I define a barrier as any policy, procedure, reward, punishment, attitude, and/or method of personal interaction that inhibits people from thinking, talking, working, and acting in support of the new management system. Eight major barriers keep cropping up:

1. Lack of commitment by top management;
2. Concerns about the ways people are judged—performance appraisals, merit ratings;
3. Poor communication or poor treatment;
4. Employment security concerns;
5. Inadequate training;
6. Resistance by unions, due to the threat of lower employment levels;
7. Uncertainty about what the advantages are for the individual;
8. Uncertainty about how to get started and what to do.

The order of importance, of course, varies by organization. At some companies concern about employment security is the most serious barrier, while others have major problems with their performance review and pay systems. But the one that usually heads the list is lack of commitment by top management. Changes in the human relations system which address people's concerns go a long way in proving top management commitment. The new system drives out fear and opens up lines of communication. It shows that people are recognized, rewarded, and promoted on the basis of new system behavior. It ensures that people get the training they need to work successfully in the new way. It helps unions understand that **waste is the most serious cause of employment insecurity.**

As a company begins to work in the new way, typical reactions are that nothing is happening fast enough, that leadership is lacking, that training is inadequate. A major culture change never goes smoothly. Many people sit back and wait to see if the changes will just go away and be replaced by some other effort.

Leaders need to understand that in making a cultural change, their actions do in fact speak louder than their words. People are constantly observing management's actions and reflecting such actions in the way they think, talk, work, and act. Leaders not only have to **say** that things are different, they have to **prove** it through their actions, through the organization's written and unwritten policies, and through how those policies are put into practice. The systems of rewards and recognition, as well as the individual behavior of the managers, should reinforce the goals of the new management system.

It is usually difficult to specify in advance exactly what changes need to be made in an organization's human relations system. Every organization has a different culture, a different way of working. The organization should change gradually to eliminate the barriers to continuous improvement as these barriers become apparent. Changes to major human resource systems like performance appraisals and pay need to be phased in over years and should include reviews and studies at various levels to ensure effectiveness. The organization can, however, begin immediately promoting people who are working in the new way.

This approach has the advantage of showing people a direct causal relationship between the change in the human relations system and the change to the new way of working. Management can be constantly looking for opportunities to make visible changes, which then serve as both proof and reminder of their commitment to working in the new way. The goal is to get people to become part of the team, to want to share in the success of the organization, to recognize that only through the success of the organization can they themselves be successful.

Five important aspects of personal interaction in the new system help to create the team atmosphere which keeps all people working toward the same goal:

1. Communicating;
2. Showing interest and appreciation;
3. Handling mistakes professionally;
4. Enabling, empowering, and entrusting;
5. Team building.

COMMUNICATING

People will believe that they are integral parts of an organization only if they feel that communication goes in all directions—up, down, and across all levels of the organization. Poor communication is one of the greatest barriers to continuous improvement. People know that communication is a two-way process—they want to see that managers are listening to what people have to say as well as relaying information.

Information, like knowledge, is power. Many people mistakenly believe that sharing information somehow diminishes their position. In fact, sharing is the only way that people can ensure that they are building a relationship based on trust.

Trusting comes in part from sharing the bad as well as the good news. Managers need to inform employees as problems that affect them occur or are forecast to occur so that no one is caught off-guard. And if things are going well, the organization will benefit from telling everyone. Otherwise people may believe that good things happen purely by chance.

People particularly need to hear news about employment security, a primary concern of most employees. The organization must inform them about all matters that may affect their job—or that they perceive may affect their job. Managers should discuss hiring policies and potential problems or changes in the business. They must answer all questions honestly and as completely as possible. They should talk about opportunities for people to get ahead and ask about their goals and aspirations.

Listening To find out what is really going on, how people feel about things, why they feel that way, and what the problems are, managers must listen. By listening—and by paying attention to employees' charts—managers can convince employees that they really are regarded as the experts.

Listening requires one's total attention and sincere interest. People cannot feign interest and expect others to communicate openly with them. To make sure that clear communication is taking place and everyone is absorbing the same messages, people must become active listeners, "reflecting" or repeating messages back to the sender, then summarizing and confirming understanding. Since listening is a learned skill, people at all levels should be trained in effective listening as well as in all other aspects of effective communication.

Using Simple Charting Techniques One of the important advantages of using charts as tools is their value as communication devices. Charts turn data into information and make it easy to transmit that information to others and to discuss it in a meaningful way. Their use also tends to structure a discussion into meaningful channels. Pareto charts help set priorities. Fishbone charts and flow charts help everyone understand the structure of a process in the same way. All the charts quickly and dramatically define problems and opportunities. They become a second language to communicate facts and data, stripped of preconceived ideas and judgments.

Observing and Inquiring Effective communication requires careful observation and useful questions. "MBWA," Management By Walking Around, is a popular approach to moving among the workers and seeing how the work is going. However, although walking around and being friendly are good human relations techniques, the new management system requires much more. The key is what we say, observe, and do when we are walking around. In the NBC TV white paper, "If Japan Can, Why Can't We?," Dr. Deming said that one of the biggest problems in American industry today is that managers do not know where to look for waste. "They don't even know what questions to ask."

Asking the right questions is based on intelligent observation and imagineering of new ways to approach a process. Good questions probe for completely new approaches, rejecting the assumption that the way things are being done is the only way or is necessarily the best way. Often a company can gain a competitive advantage by completely shifting its approach to producing or marketing a product or service. Questioning is a difficult art, but an extremely valuable one to develop. In fact, it can sometimes make the critical difference between the success and failure of an enterprise. I will discuss it in more detail in a later section.

Managers should seek out, understand, and use variation. They need to gather the kind of information that comes from being knowledgeable, observant, and keenly interested in people and work processes. They should show that they are dissatisfied with the status quo by constantly initiating projects to fix the processes and keep them fixed. In this way, they can lead the drive for change.

SHOWING INTEREST AND APPRECIATION

In the new management system, it is important for managers to go out of their way to show that they recognize the importance of individuals and their contributions to team objectives. They can do this in many ways. Managers should express an interest in the well-being of others, in their personal life as well as their work. Some managers avoid talking to people about their personal life and problems. They should recognize that problems in one's personal life as well as one's work life can affect job performance.

A manager should also understand people's goals and aspirations. By taking a personal interest in people, managers can help foster a sense of loyalty. Of course, the interest must be sincere and non-intrusive. If employees trust their managers, they will often volunteer what they wish to discuss.

Managers should go out of their way to say thank you and to give people credit. They look for opportunities to recognize the value of people's contributions to the new management system. They express appreciation to project teams for a job well done. Notes of appreciation, verbal thank you's, visits, spontaneous conversations—all of these add up to constructive personal relations.

In my career I have received five handwritten notes from my boss, indicating he appreciated the job I was doing. These notes really pleased me, and I was proud to share them with my family and friends. They were so important to me, in fact, that I still have them—in my top dresser drawer, tucked in alongside the socks and handkerchiefs. Think of how much energy and dedication you can generate by taking the time to show honest appreciation of someone's effort.

HANDLING MISTAKES PROFESSIONALLY

Always admit mistakes up front. One of the easiest ways to get people to talk about problems and improvements is to admit that you are a part of those problems. Take the blame for mistakes that you had anything to do with. This is often the first step in granting amnesty. By acknowledging your own mistakes and giving yourself amnesty, you demonstrate that you are sincere about granting amnesty to others for their past mistakes. Forgiving and forgetting the mistakes of the past shifts the focus to improving the current work process.

Do not confront people with their mistakes without first giving them a chance to admit that they have made them. When someone does make a mistake, rather than make an issue of it or try to humiliate the person, ask a question to see what that person thinks could have been done differently to avoid the mistake. Probe how the system could be improved.

Take the following situation: You asked a member of your staff to organize a meeting for you and several things went wrong. You were given the wrong room number and arrived at the meeting late. The person in charge of having supplies for the presentation forgot the spare bulb for the overhead projector, and the presentation was delayed when the bulb burned out.

Nothing would be gained by launching a verbal attack and reminding the person of his shortcomings. You would only be venting your frustration by making him feel badly. Instead, by calmly asking the person how he felt the meeting went, you will likely elicit an acknowledgement of the problems and some suggestions as to how they can be avoided in the future.

By showing that no one is going to be attacked for errors, you support openness in the new management system. Your behavior reinforces the concept that most problems are the fault of the system and not the fault of the individual. It shows that you are serious about fixing the system or processes that you control, and it encourages others to adopt the same attitude.

ENABLING, EMPOWERING, ENTRUSTING

In the new management system, managers at all levels are responsible for enabling people to work to the best of their abilities. Few functions are more important. Managers show respect for employees' ideas and empower them by providing the resources so they can carry out projects for continuous improvement. The expert workers should be entrusted with leadership for some projects, a role shift that can generate a complete change in attitude along with tremendous enthusiasm. The feeling is "Wow! They're letting us do what we know how to do!"

In the old system, managers tended to tell people what to do, to direct activities, and to judge whether the people carried them out. In the new system, managers are coaches who set an example and provide suggestions. They give workers the power and responsibility to make continuous improvements in an atmosphere of mutual trust.

TEAM BUILDING

Team building is one of the most important aspects of working in the new way. Individuals who identify with a team align their personal objectives with those of the organization. Nothing is more important for the success of an enterprise than having all the people in it supporting it and working to achieve its objectives. Many Japanese companies have done an outstanding job of getting all their people to work for the good of the whole organization. Dow Chemical prides itself on its groups that bring together people from any area and any level to work together. These groups work towards a common goal, sometimes on a temporary, ad hoc basis, but usually on a more permanent basis. Teams can work this way without undermining respect for authority; in fact, teamwork generates mutual respect.

So far, Dow's groups are the exception, rather than the rule, in the United States. Western society, especially American, is very individualistic. However, many Americans enjoy being part of an athletic team. If that dedication to the team can be transferred to the workplace, individuals' objectives will be aligned with those of the organization, and the organization will be fully using the dedication and brainpower of all its people. With this base, the only additional ingredients necessary to approach the imagineered perfect human relations system are training and the coaching provided by management.

Emphasizing teamwork takes the emphasis off the individual. When problems arise, the focus is finding the cause of the problems instead of finding the scapegoat. Likewise, the group, the team, receives the organization's commendations. Organizations need project teams made up of people from all appropriate areas to study thoroughly and improve continuously the organization's processes and systems. When teams, working together to solve a problem, make a decision, the team members buy in to that decision. Then they work together to fix the problem and keep it fixed.

An emphasis on teams does not mean that individuals lose their importance. Organizations continue to acknowledge and reward individuals' superior achievements when those achievements are exceptional (out of the system) and accepted as such by the team. But when the team has done the work, the team should get the credit.

Teams working toward a common objective generate a lot of power and a spirit that outlasts the objective and continues to produce dedication and enthusiasm. In May of 1982, the Computer Products Division of Nashua Corporation was growing rapidly. However, it was having difficulty matching orders with production capabilities and couldn't seem to get beyond a plateau of sales volume. The division broke through that barrier with the aid of T-shirts, of all things. Several supervisors appeared one day wearing T-shirts saying "$9,000,000 in June." Immediately everyone wanted one, and the whole division became obsessed with working together to achieve that sales goal. Sales people asked manufacturing what type of orders they could most easily produce, and manufacturing asked salespeople what products were most in demand. With such communication, the division met its objective.

To recognize the achievement, a barbecue was held for the entire division. The food was served by the division general manager, the vice president, and other division management. Everyone felt a great sense of achievement and saw that others recognized their efforts. The team spirit generated by achieving that objective carried over and provided the impetus for further months of record sales.

Chapter 11
LEADERSHIP

Implementing substantial change requires a leadership style seldom found in American corporations. Management by planning, budgeting, and controlling does not usually lead to major changes in the way organizations operate. Transforming the way people think, talk, work, and act requires active leadership.

In the old system, the manager analyzes, makes decisions, tells employees what to do, and then judges their performance. In the new system the manager-leader acts as a player-coach. This participatory management is very different from the typical authoritarian, bureaucratic style still dominant in American organizations. As we discussed in the last chapter, people will use all of their energy and brainpower only if they feel like important players on an organizational team.

PROMOTING THE VISION

Leaders in the new system promote—perhaps even "sell"—the organization's mission and vision, topics covered in detail in Chapter 15. They work to get everyone's support for the vision, help determine what needs to be done to achieve it, and energize the team(s) to do their part enthusiastically in making the vision a reality.

Leaders must be **enablers** for the team(s). They help provide the necessary resources, including education and training, and help remove any barriers.

Leaders **empower** others, giving them the authority and the power to effect change themselves and to lead others into the new management system.

Leaders also **entrust** others. They recognize that people need to be shown that they are trustworthy—so that they will want to demonstrate by their performance that they deserve that trust.

Leaders need to be trustworthy themselves. For major change to be made, people must trust the leader who is asking them to change. As Edward R. Murrow said, "To be persuasive we must be believable; to be believable, we must be credible; to be credible, we must be truthful."

Leaders in the new system not only encourage people to take initiative, they also actively promote change. Change means uncertainty, and for people to want change, they must have a vision of the better future this change will bring. Leaders communicate the vision of the organization at every opportunity. Everyone will not understand or accept the vision at first. Repetition, discussion, asking and answering questions helps. Prominently posting charts of the operating variables that support the vision is a powerful message and constant reminder of what the leaders consider important. But leaders constantly go beyond communicating by word and chart and show by their actions that they enthusiastically support the vision. Nothing will do more harm to the process of cultural change than people in positions of authority merely giving lip service to the need for change.

Leaders show their support for the vision by being active, up front, visible participants in making changes. While they delegate, they also participate. They coach and advise the team, but they also play with the team. They lead by word and by example; they should be seen using the tools of continuous improvement, leading projects, and teaching others at every opportunity. They know in enough detail what their team(s) are doing so they can coach and ask questions intelligently. They use the principles of continuous improvement while making their most important decisions. They always keep the vision in mind and keep developing support for the vision throughout the organization. All the leaders of change share this constancy of purpose. A successful atmosphere of change generates a myriad of projects, large and small. The project teams recognize that even the small ones support the vision.

Leaders' work to align people in support of the vision continues outside the organization. Leaders elicit support for the vision from key suppliers, customers, and others affecting the operation. Because the vision by definition includes serving customers better, involving customers in the vision is a good marketing program. If the organization is part of a larger organization or is influenced by other groups, these constituencies also must be educated to share the vision.

Leaders undertake a number of activities to move the organization toward continuous improvement and toward achieving the vision. They help to choose projects and project teams, and they allocate resources to those teams. In making allocations they maintain a Pareto mentality and give most support and priority to those projects most important to achieving the vision. They observe and/or participate in team meetings periodically to help determine when additional help or perhaps a different direction is needed. They try to ensure that the group truly has a team orientation and that all members are working toward the common goal. They keep people focused on improving the process and not on finding who is at fault.

Leaders are risk takers and encourage projects that may not have an assured payoff. Leaders may therefore have to accept setbacks, but effective leaders view these setbacks as learning experiences, focusing on gains and not losses. They are aware of potential losses but constantly drive, push, talk, and act to inspire everyone to strive for the gains.

LEADERSHIP DEFINED

Because of the importance of leading by example, I define leadership as "the visible, observable behavior that makes people want to follow or emulate another. In the new management system, it is the behavior that indicates a willingness to take whatever steps are necessary to find, quantify, and eliminate the waste and work toward continuous improvement."

As this definition makes clear, a leader does not have to be a manager. In fact, in the new system leaders in continuous improvement spring up throughout the organization. Machine operators or workers on the assembly line who diligently and enthusiastically work to eliminate waste can serve as role models and leaders for their peers, who may seek to emulate them.

Management recognizes and supports such leaders of continuous improvement, whatever their position in the organization. This support encourages others to follow their example. The organization should gradually change the formal and informal reward systems so that people perceive that it is in their interest to work in the new way. In fact they should eventually perceive that continuous improvement is the only acceptable way to act in the organization.

Getting people to want to change the way they think, talk, work, and act is a major accomplishment, but it is not sufficient. Attaining a vision that necessitates substantial change almost always requires tremendous energy to overcome barriers and sustain the effort over a long period of time. Leaders generate this energy by changing people's basic attitude about their work. Economic incentives are not enough. They must appeal to people's basic self esteem. Fortunately the new system of management makes that appeal. People are now asked for their ideas and have more personal control of, and impact on, the process in which they are working. Functioning as a valued member of a successful team powerfully enhances self esteem. Leaders ensure that the enabling, empowering, and entrusting are real, and that the team efforts receive recognition and appropriate rewards, creating pride and again enhancing self esteem.

Beyond this the leaders provide that energy that comes from their drive and enthusiasm for continuous improvement, making it even more intense when obstacles occur. High energy can be contagious and can inspire extra effort in others. Just as an athletic team needs that extra effort in a crisis to overcome a tough opponent, so an organizational team needs periods of high energy and extra effort to overcome barriers on the path to attaining its vision.

To become a world class competitor in today's economic environment requires this kind of leadership and focus on continuous improvement. Charisma is not required. Instead, leaders need a strong will to make the changes, the belief it can be done, the education and training to provide the wherewithal, and action.

Chapter 12

ORGANIZATIONAL ASPECTS OF THE HUMAN RELATIONS SYSTEM

The success of a human relations system for continuous improvement depends as much on the organization's policies and procedures as on managerial behavior. People cannot work in the new system without the full support of the organization.

Through policies and procedures, an organization determines such things as pay scales, supervisory levels, and promotions. These policies and procedures play an important role in determining people's attitudes toward their work and the organization. The essential organizational aspects of a human relations system are rewards, advancement, recognition, and education. Each element should reinforce the goals of the new management system.

Again, managerial actions speak louder than words. It is not enough to say that people should work in the new way. Written and unwritten policies and procedures should demonstrate that working in the new way is best for the individual as well as the organization. Through its human relations system, both formal and informal, management shows that it is sincere about wanting people to think, talk, work, and act in the new way.

Organizations should strive to create a human relations system in which individual goals are consistent with the organization's goals. To do this, managers establish a system of rewards, advancement, recognition, and education that eliminates conflicts between what is good for the individual and the team and what is good for the enterprise. At the end of this chapter are five questions against which people should regularly test all personal and organizational behavior.

REWARDS

Both monetary and non-monetary rewards retain their importance in the new system. Monetary rewards include pay, bonuses, stock gifts, and merit raises. Non-monetary rewards include special prizes, dinners, plaques, tickets to events, and all the other ways organizations can show their appreciation for an individual or a team without fattening anyone's wallet. The effectiveness of the traditional human relations tools, including non-monetary rewards, should not be underestimated.

ADVANCEMENT

People learn what the organization wants by observing the behavior of people who get promoted. Therefore, managers who are serious about working in the new way promote workers who have joined the team in the battle against waste, and managers who are most committed to continuous improvement climb the management ladder most quickly. If people see that striving for continuous improvement is a major criterion for advancement, they will quickly realize that it is in their self-interest to work in the new way. And if people understand that teamwork is expected as part of their job performance, they will more readily share credit for accomplishments.

RECOGNITION

People want their effort and contribution recognized, and that desire is a powerful motivator. Non-monetary recognition can be just as important as monetary recognition, and it more directly supports teamwork. Too often, management recognizes accomplishments with money when what people crave is a word of praise.

To give non-monetary recognition

1. Say thank you;
2. Write thank you notes;

3. Make sure top management hears about progress working in the new system;

4. Arrange a presentation to top management of a successful project by the participants;

5. Arrange for write-ups in the company newsletter;

6. Include success stories in the annual report;

7. Distribute case studies on successful projects throughout the organization;

8. Send out news releases on successes;

9. Arrange for awards of some sort to recognize those project teams that have been successful in making significant improvements in a process;

10. Arrange for broad publicity for award winners, making sure that their families and friends are aware of the importance of their achievements. A newsletter sent to the employees' homes is a good vehicle.

EDUCATION

Education (the "know-why") and training (the "know-how") are important human relations tools for overcoming barriers that prevent the new management system from being implemented successfully. People must be taught why working in the new management system is central to their self-interest and the good of the organization. For instance, if people know that working in the new way enables them to help anyone or any group, team, or organization to raise its performance substantially, most would change much more quickly. People working in the new way can help families and friends to do almost anything more effectively. People must also be trained in new management tools, techniques, and concepts. Education lets people understand why changes are needed and therefore adds to their motivation.

For education and training to be effective, they must be integrated into all phases of the business. For example, many companies offer courses on technical tools, statistical control, and participative management. Unless and until those courses are tied directly to working toward continuous improvement, they lose a great deal of their value.

People have to understand why they are going to a participative management or leadership course, why they need training on such things as simple charting techniques, communication, and work sampling. They need to understand that they are participating so they can help work successfully on projects to find waste, get rid of it, and prevent its return. They are participating so they can help their organization and the people they work with make continuous improvement in all areas of the business.

The courses also have to be timely. Employees will forget statistical and charting techniques if they do not use them within a few months. The training for each group should be timed so that project work will immediately follow the training. The people who will be working together on a project should be trained together.

In some cases "just-in-time training" can be carried further. The project leader can wait until he or she sees the need for a particular technique and then train the team in that technique, e.g. histograms.

PERFORMANCE REVIEWS

As they are traditionally used, personnel appraisals, including performance reviews and merit ratings, present problems in a system of continuous improvement. Dr. Deming views them as "one of the deadly diseases" afflicting Western-style management. And with good reason.

Such appraisals tend to focus on results, generally short-term, based on some sort of verifiable count that most often, as Dr. Deming puts it, robs people of their pride of workmanship. For example, an individual's performance ranking is sometimes based on how many units he or she can produce in a given time period. Taking the time to examine the process or system in an effort to improve the quality of those units might adversely affect the individual's rating. The individual therefore is not encouraged to work to improve the process. These kinds of rankings also build fear: "If I question the work process, my boss will think I am challenging him and won't give me a good rating."

In the new management system, performance reviews shed the focus on short-term outcome. To maximize the benefits of performance reviews or appraisals, management applies the principles of the new management system, including the concepts of variation.

Random variation exists in individual performance just as it does in every other system. The random variation cannot be sorted out. Instead, what stands out in individual performances is the non-random variation—the exceptional people at either end of the job performance scale. There are usually exceptional people who do work far superior to almost everyone else's. And there are people whose jobs should be changed because they lack the basic skills for that particular task. However, most of the people—well over 90% in a typical organization—will be "in the system." Unfortunately, performance appraisal systems often ask managers to act on random variation, to sort people who are part of the system into categories. For many reasons, people regard the appraisal system as unfair.

Over time, one thing should be made clear in reviews and appraisals—that working in the new system is the only acceptable behavior. If a basketball team is operating a zone defense, it cannot have one player playing man-to-man. For teamwork to be effective, players cannot operate under a different system than the one spelled out by the coach. It must come across strongly that the organization is following the new system, and any members who want to remain on the team are expected to work in that system. Managers—coaches—work to improve the skills of individuals, but more importantly they strive to shift the system and raise the overall level of performance.

MBO

Management by Objective (MBO) is another technique used by many organizations today in an attempt to measure and improve individual performance. In MBO systems, managers work with their bosses to set objectives and are measured by their ability to meet those objectives. Unfortunately, the practical result in too many cases is that MBO sets de facto caps on work standards and on quality and productivity.

The reasons are clear. When setting objectives, people are naturally going to make them as easily achievable as possible to make sure they can meet them, particularly when they know that punishment will follow missed objectives. Otherwise, they risk failure, which affects their performance review, which in turn affects their chances for pay increases and promotions.

What's more, they know they will be expected to top those objectives every year or every review period. So, even if people know they can comfortably exceed the objectives, they logically aim low and gradually work up to the level that they already know is possible. MBO contributes to a fixation on short-term results rather than on long-term system/process improvements. It also assumes that individuals control the process or system, which is not true most of the time.

In continuous improvement, MBO should be replaced by management with the single objective of continuous improvement. This is an objective that everyone can understand and accept. It does not set caps on quality and productivity. Nevertheless, a "quality performance index" can be useful in measuring and motivating a group or a team. For a manufacturing group, this index might include quality of products or services, productivity, safety, and absenteeism. To be useful, it should be devised and agreed upon by the group being measured. The group will then look on it as a way of determining progress toward continuous improvement, and it can thus become a motivator.

MBO should become MBIP—Management by Improving the Process. As part of the organization's business plan, employees and managers should agree on minimum objectives that both parties are confident will be exceeded. They then work together as a team. Every four to six weeks—not once per year—the boss reviews with each employee the team's performance and that employee's performance as part of the team. In such a system, the boss is the coach and leader, the person who helps raise the average, the one who eliminates obstacles others cannot remove. Everyone works together as a team to improve performance continually—well beyond the objective.

HUMAN RELATIONS QUESTIONS

Here are the process checks for personal and organizational behavior. Test all of your human relations against the following five questions:

1. Do you treat people as you would like to be treated if you were in their position? Suppose that in every decision affecting people, you raised and examined that question. How many decisions dealing with people would you make differently?

2. Do you recognize that 90% or more of all the troubles, waste, and opportunities for improvement come from the process or the system, not the individual? How often do you forget that fact and blame or praise the individual? "John did it." Often what happened was that conditions outside of John's control changed. Good or bad, more than 90% of the time it is the system, policies, and/or process. That is why work on process improvement is so critical.

3. Do people in the organization perceive it is in their interest to share their knowledge of the best way to work, to cooperate rather than compete? The work may be selling to major accounts, developing the strategic plan, handling customer orders, repairing a building or machine, designing a new mold, or operating a particular machine. If the experts don't see that they benefit from sharing their knowledge, how are you going to bring about continuous improvement and constantly raise the average performance of the entire group?

4. Do you have artificial caps on the quality and productivity of the work? Historically, those caps were work standards, used in the 1920's, 1930's, and 1940's for workers in repetitive operations in manufacturing plants. Currently, management by objective often does the same thing with managers, supervisors, and independent contributors, encouraging people to set low objectives. If people are fearful that they have to meet every objective, they are likely to set objectives they know they can meet. Does that happen in your organization?

Many organizations also use their budgets as a cap. Every business has budgets or plans, but if the budget sets acceptable levels of work, then it too may undermine the drive to make improvement a continuous process. Budgets should not set work standards.

5. Do people in your organization know how the new system benefits them? No matter what they are told, people generally do what they want to do, and most people want to do things that

are in their best interest. If they don't perceive changes to be in their best interest, they won't give them their enthusiastic support, even if they are convinced that the changes help the organization.

If you test your policies, procedures, and rules with these five key questions about human relations and make changes accordingly, you can avoid many of the traps that prevent or reduce continuous improvement.

Chapter 13

CUSTOMER AND SUPPLIER RELATIONS

In the new system of management, relations with customers and suppliers are an extension of interpersonal relations within the organization. People within the organization strive to treat customers and suppliers as part of a team rather than getting into the "us versus them" mentality common in traditional management systems.

The new management system is strongly dedicated to pleasing the customer. As Dr. Deming said, "No customers, no orders. No orders, no jobs." To please customers, an organization needs to understand their needs and wants, whether the customers are processes, individuals, or other organizations, internal or external.

Sometimes the customers do not know what they want or need. They may not understand or be aware of the current and future changes in the market or in the technology. In such cases, an organization may need to work with its customer to help it prepare for such changes. Remember that customers cannot tell an organization that its costs are too high or productivity too low. The organization has to make such judgments itself through knowledge of the work, technology, and benchmarking.

COMMUNICATING WITH CUSTOMERS AND SUPPLIERS

Charts can help improve communication with both customers and suppliers. Suppliers and customers are a part of the organization's work process, and it is important that they understand each other's capabilities and requirements. The dynamics of customer and supplier relationships can be shown as follows:

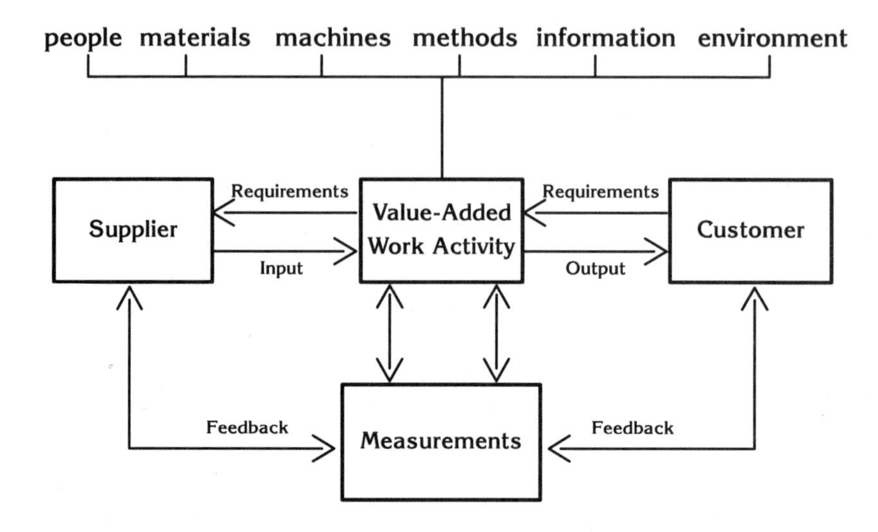

In the old system the major means of communication between suppliers and customers were specifications, prices, and complaints. Suppliers often hid problems from customers. In the new system, suppliers and customers work as a team. Each tries to understand the other's capabilities and requirements with the objective of maximizing quality and minimizing overall cost. Supplier problems are looked on as problems of the overall process. Suppliers and customers work together to solve them.

A company's record of complaints provides one measure of how successfully it pleases its customers. Categorizing the complaints and making a Pareto chart of the reasons for the complaints can reveal some of the things the company should be improving. The most devastating complaints, however, the company never hears; the customer says simply, "I'll take my business elsewhere." Often companies need to seek out their customers and ask how they feel about the relationship. Organizations need to look at complaining customers as highly valuable assets; they can lead the organization to problems and waste and perhaps prevent the loss of scores of unhappy customers that the organization never hears from.

The local Toyota service department calls all its customers the day following any service and asks a series of questions to determine the degree of customer satisfaction. The answers provide valuable feedback and the call itself often serves to defuse any antagonism that might lead to the loss of a customer. The process gives Toyota a definite edge over its competition, who judge how they are perceived by their customers solely on the basis of the complaints they receive.

CUSTOMER-SUPPLIER TEAMS

Insofar as possible, customers and suppliers should work as a team. When you solicit a customer's help in doing a better job, he or she will be inclined to join your team. If you are so fortunate as to have customers who are working in the new management system, your ability to provide statistical information about your product and your work on continuous improvement will give you a decided edge over competitors still working in the old way.

You should certainly be looking for suppliers who are working in the new way. In the old system, each shipment had to be inspected to determine if it was acceptable. Even if the inspected sample was satisfactory, the uninspected majority of the shipment could cause problems. Some companies reship returned goods, hoping that the next customer will find only acceptable goods in its inspection sample. Customers feel much more confident if with each shipment they receive copies of control charts of the critical processes and histograms showing the distribution of the critical factors. Assuming you have come to trust the supplier, you don't need to inspect the shipment, and you know exactly what you are getting. Many companies now have vendor certification programs that allow them to accept products on the basis of statistical documentation, rather than by inspection. Think of the time and money saved. Of course the biggest reason to seek out suppliers working the new way is that they are likely to have cut out most of the 20–50% waste that companies buy from traditional suppliers.

Working closely with a supplier can also save the time and money wasted on over- and under-specification. The more you and the supplier understand each other's capabilities and requirements, the better you can tailor the specifications to lead to lower overall costs. Sometimes a slight change in specifications that means little or nothing to you will allow a substantial reduction in the supplier's costs. Likewise, a supplier who understands your process may be able to make some slight changes in the product that will reduce the cost of your process.

Working with suppliers this way has an important implication. You must not have too many of them. Every supplier, even within the same company, will have different process variations. It is a time-consuming process for each of you to understand the other's capabilities. You can't afford to spend the time on multiple suppliers for most things. For each item, you need to find one supplier who will become part of your team and establish a good, long-term relationship. Only as a team will you be able to make continuous improvement throughout all parts of the process. A good supplier who is part of the team can also be brought in on new product development, so that the new process can be optimized for lowest overall cost.

Advice on choosing suppliers is included in Dr. Deming's famous 14 points. His point number four warns "End the practice of awarding business on the basis of price alone." Looking only at a supplier's prices is a very short-sighted practice that ignores quality, service, uniformity, reliability, and the advantages of working with a supplier as part of a team on a long-term basis. Unfortunately, many purchasing people are trained and judged on the basis of always getting the lowest price. In the new system they are trained to pick suppliers that allow the overall lowest cost and best quality for the end product. They are also trained in charting techniques so they can communicate with the supplier accurately about product characteristics and process capabilities.

Many of the people in an organization deal only with *internal* customers. While pleasing the external customer must always be the foremost objective, most of the same principles apply to suppliers and customers within an organization. In the old system the various departments have walls around them. People are concerned with meeting the objectives of their own departments. In the new system people try to optimize the work of the whole organization, and they function as teams. The old style organization looks like a series of silos, with the only communication between groups occurring at the top of the silos. The new way rips out the walls and encourages horizontal communication and cooperation between supplier and customer departments.

When dealing with internal customers, always investigate whether requests for work are also in the interest of the external customer. If they are not, pleasing the internal customer may create more waste. This kind of waste is quite common in most large organizations.

SECTION IV

CREATING THE NEW SYSTEM

We end with the beginning of the continuous improvement process because we hope that by this time in the book you are beginning to recognize the power of continuous improvement and are ready at least to probe the possibility of adopting The Right Way to Manage. As should be clear by now, the new way is not a club that an organization can benefit from simply by signing up and paying dues. It requires a serious commitment from the organization and the individuals within it. If we had put "Creating the New System" at the beginning, enthusiastic managers might have attempted to undertake the transformation without fully understanding what it involves or learning to use its tools. Such attempts inevitably end in frustration.

Having made it this far in the book, however, you now have some sense of what you're getting into. If your enthusiasm has been fueled by stories of other organizations' successes and by your vision of how your own organization can change, you're ready for the next three chapters. Chapter 14 briefly outlines what an organization needs to get started in the new way. Reading it should let you evaluate how ready your organization is and what changes are needed before the transformation can begin. Chapter 15 summarizes how an organization can actually start to take action. And Chapter 16 presents some questions that you can start asking yourself now and that may become a guide for you as you proceed in adopting the new way.

Chapter 14
GETTING STARTED

To accomplish anything, you need

1. the will to do it,
2. the belief that it can be done,
3. the wherewithal to do it, and
4. action.

If you don't want to do something, or don't believe it can be done, or don't have the tools to do it, it won't get done.

THE WILL TO DO IT

Of these four elements, the one that organizations most often lack is the will to do it. Lack of will keeps organizations from adopting The Right Way to Manage. They lack will because they sense that the new system requires a change or *shift* in the model or *paradigm* on which people base their thinking. A *paradigm shift* involves replacing the accepted model with a new paradigm.

The classic paradigm shift was the change from "the earth is flat" to "the earth is round." This was not an easy change for people to accept. It went against what their observations and common sense told them. It required a significant, dramatic change in their thinking. The new system of management involves a similar shattering of theories, a significant alteration in the way we think about management and the way we work. It is difficult to accept that the way we have always done things is wrong. As Daniel Boorstin said, in *The Discoverers,* "The greatest obstacle to discovering the shape of the earth, the continents and the ocean was not ignorance, but the illusion of knowledge."[16]

Illusions persist in management theory and practice. Elegant theories of management that concentrate on financial and marketing strategy and ignore the imperative to make continuous improvements in operations have gained followers in the United States and are taught in most American business schools. In 1984, two distinguished business school professors, Robert H. Hayes of Harvard and Steven C. Wheelwright of Stanford, summarized the effects of the dominance of these management theories.

> *Pursuit of these modern management approaches has led American manufacturing companies to:*
>
> *1. Emphasize analytical detachment and strategic elegance over hands-on experience and well-managed line operations.*
>
> *2. Focus on short-term results rather than longer-term goals and capabilities.*
>
> *3. Emphasize the management of marketing and financial resources at the expense of manufacturing and technological resources.*[17]

People trying to change the approach to work in their organizations often must overcome the results of years of business school education with this kind of emphasis. Changing our way of thinking about what is important in managing an organization requires a leap of faith. Sometimes people will make this leap only if they recognize that the present system is not working.

People also resist the new philosophy of continuous improvement and continuous change because they resist change itself. The self-preservation instinct tells people to maintain the status quo. Making change a major goal requires a radical shift in thinking. Many people and organizations embrace change only when the status quo is no longer viable, or when they see that change will improve their lot.

Because people are so likely to resist change, it is critical for them to understand the desirability or necessity of changing. Managing the new way requires not just a change in behaviors but a change in philosophy; it involves different beliefs, concepts and attitudes. It is a *pervasive* change.

Management must, therefore, sell the change and perhaps even require it. To do so, managers themselves must first understand and believe in the value of the new philosophy. Top management becomes educated about the way this philosophy works and how effective it has been in other organizations. Only then can management provide the necessary active, visible, up-front leadership of the change.

Why should managers attempt to make a major cultural change in their organizations? Almost 500 years ago, Machiavelli warned in *The Prince* that "It must be remembered that there is nothing more difficult to plan, more doubtful of success, nor more dangerous to manage than the creation of a new system. For the initiator has the enmity of all who would profit by the preservation of the old institution and merely lukewarm defenders in those who would gain by the new one."[18]

There are two basic reasons—one positive and one negative—for overcoming such enmity and undertaking this major change. From the positive side, it enables organizations to please their customers through quantum leaps in quality, consistency, and reliability, while at the same time reducing costs dramatically. The negative reason is simple. If an organization doesn't change, and its competitor does, it may find itself out of business.

In the computer memory disc operations at Nashua, this change was a tremendous help in improving our competitive position. We had been working in the new way for about nine months with the dramatic increases in quality and yield I outlined earlier. We decided to use this success story to sell more product. We went to one of the most demanding users of memory discs in the business—Hewlett Packard Computer Group in Boise, Idaho. We had never done any business with them before. We told our story to their top people, and they were sufficiently impressed that three of them agreed to come to Nashua for a day, to see for themselves. They came and stayed three days instead of one. They were so amazed at what they saw that at the end of the first day, they agreed to give us all their disc business over the next six months.

We could never have won such a contract or such respect just by giving a low quote. What they wanted was a supplier with a process that would insure them of a consistent high-quality product that met their needs. They liked the idea of developing future projects with a supplier that was continually improving its process capability. We worked together during the design phase of a new product so that we could tailor a memory disc to their needs. No one competed with us for the business because Hewlett Packard was convinced that our working method would get them the product they wanted at prices they were willing to pay.

The hard disc business also illustrates what can happen when you don't work this new way and a competitor does. I can tell this story about Control Data because a vice president of that company told it himself when introducing me at a supplier seminar in 1988. He had seemed anxious to talk to me privately at one of my seminars in Nashua, NH, about a year prior to that. As I recall, he came up to me after the first session and said: "Bill, tell me. It must have been a real breakthrough in your disc business. It must have been a technological breakthrough, to put all those companies out of the business. Divisions of the best companies in the world going out of business in just a few years."

"I watched you, Bill, as they all went out of business. You know what happened to us, don't you? I took over in '81 as head of Control Data Peripherals, the most profitable part of CD, with over $1 billion sales. The most profitable part of all that was the hard memory disc business. You know what you did to us, Bill. You put us out of business. At the end of 1985 we shut the doors on the disc business, sold all the equipment and inventory for 10 cents on the dollar."

It *was* a technological breakthrough that allowed us to put other companies out of business, but not the type that he originally thought. It was a breakthrough in the technology of management.

The other companies he referred to were BASF in the U.S., 3M, Xerox and Memorex. None of them could match our quality and cost, even though they were all known for quality products.

Working hard in the conventional way is not good enough when a competitor adopts the new way. Remember, all the changes at Nashua came from changing the way we worked. All the people involved were the very same ones who had made us an average competitor prior to Dr. Deming's arrival. It was not the people, it never was . . . it was the management system.

Since so many changes in basic attitudes, concepts, and beliefs are required, nothing will happen unless management has a strong desire and will to make it happen. Because many people cannot summon the will to make such changes, the new philosophy has been slow to spread in the Western world. Ironically, Japanese businesses got such a head start in adopting the new philosophy because they were in such a desperate position they realized they had to change to survive.

THE BELIEF THAT IT CAN BE DONE

In order to act upon their will or desire, people need to believe that the change can be successful. This too is a difficult step. The belief in the feasibility of the process may come partially from reading about and talking to people who have been successful. Unfortunately, total belief comes only from personal experience, from doing it yourself and seeing that it does work. So it takes a leap of faith or at least a suspension of disbelief to get started. Then, when you have worked on it long enough and hard enough to see its power, belief will come. You can tell children they can swim, and they may believe it intellectually while still standing on the shore. But they are not quite sure that it really applies to them until they actually get into the water and do it. Only the personal experience of success makes them true believers.

THE WHEREWITHAL TO DO IT

Although many people react to any suggestion of change by asking "Where are we going to get the money?," obtaining the wherewithal may be the easiest part of the management transformation. It doesn't require capital expenditures or investment in inventories or R&D. It is a matter of obtaining education and training for management and, to some degree, for everyone in the organization. Anyone who desires such education can now find it. Books like the ones listed in the bibliography at the end of this book can provide excellent initial exposure. Interested people can attend seminars or view videotapes by Conway Quality and others.

Often a company will send a "scout" to attend such a seminar to get a feeling for what is available. We usually have several scouts at each of our public seminars. This approach works only

I keep talking about the massive change in attitud to work the new way because it's not just a matter of usir cal tools in a new way. It means making all the memb organization (including suppliers and customers) part a team devoted to continuous improvement. Let's s some of the differences between the old system and many of which we have already explored in detail.

OLD	NEW
Authoritarian	Participative
Fear of Job Loss	Employment Security
Fear of Reprisal	Open Atmosphere
Position Means Knowledge/Authority	Authority of Knowledge The Workers are "The
Status Quo—Good Enough	Continuous Improvemen
Systems, Policies, Procedures for Control	Systems, Policies, Procedures for Suppor
Beat up on Suppliers	Supplier Partnerships
Quality—Conformance to Specifications/Product	Quality—As Defined by your Customers/Proce
Specialists in Statistics, Industrial Engineering	Everyone Trained in Concepts and Tools
Driven by Management, Cost, or Expense	Driven by External Custo including need for low
Individuals Rewarded	Teams Rewarded
Work on Results	Work on Process
Quality vs. Productivity	Productivity through Qua
Functional Structure	Cross-Functional Teams
Driven by Financial Performance	Driven by Quality Measur
Short-term Perspective	Long-term Perspective

if the scout holds a respected position in the company, so that his or her analysis and judgment will be accepted by top management leaders. Eventually, if the change is to succeed, the CEO—or at least the head of the division—must make the effort to become educated. The leaders of this important change cannot stand on the sidelines or delegate this responsibility. They must be personally involved as participants, coaches, and salespeople for the new philosophy. They cannot do that effectively unless they themselves have taken the time to be educated and trained. The personal commitment by the CEO in any organization is all-important. If the new system is to have a chance of success, management and leaders at all levels of the organization must understand what it can do for them and have a strong desire to make it work.

Education provides an understanding of the objectives and power of The Right Way to Manage sufficient to generate the *will* to do it and part of the *belief* it can be done. Training provides the tools that constitute the *wherewithal* to change to the new philosophy.

Managers and potential leaders of the change throughout the organization need to be *educated* so that they will enthusiastically lead the change, and they must be *trained* in the concepts and tools of variation so that they can implement changes themselves and coach others. It is not enough for managers to "support" the effort. They must be actively involved as participants and leaders, with projects of their own. They must act as leaders as well as managers.

A thorough understanding of the "know-why" must prevail at least down to second level managers like superintendents and department heads. Training is necessary at all levels, but the education-training mix becomes more and more oriented toward training the further down you go in the organization's hierarchy. The first and second line supervisors should receive the most intensive training in the tools of variation. These are the people who will field most of the questions on a day-to-day basis. Organizations should identify leaders and champions of the program and potential internal trainers, since there will be a continuing need for retraining and for training of new employees.

People at the bottom of the organizational ladder need to understand the concepts and tools because they are the experts in their process. They are a crucial part of the team. They need the tools to identify problems effectively and to suggest and/or implement solutions. An organization can multiply its problem-solving power tremendously by making everyone a problem solver and team member. This is one of the secrets of the power of this system of management. By using the brains of those closest to the problem, an organization becomes much more effective in identifying, quantifying and eliminating the waste. It is releasing the power of the people!

Besides learning from seminars, books, and videotapes, interested managers can contact a number of companies that have gone through this cultural change and may be willing to provide information on their approach. Two excellent practitioners, Dow Chemical and the Shipley Company, have videotapes describing their change to the new system. (See Appendix.)

ACTION

Once a person or organization has the will, the belief, and the wherewithal, action follows naturally. It will eventually involve virtually everyone in the organization, as hundreds or even thousands of improvement projects spring up. Initially they should be relatively small, tightly focused projects. It is important that the first few succeed—not every one, but the majority of them. Everyone will be watching, and each success will cause more people to begin to change their thinking. Later, senior management should identify and lead a few major programs of critical importance.

When we put together will, belief, wherewithal, and action, we have an engine of progress, an engine of change. When we apply this to The Right Way to Manage, I call it the Conway Engine. It looks something like Figure 14-1.

This engine drives a continuous effort to *identify, quantify and eliminate waste through process improvement.* This is the core activity of all the people in the organization. It will be accomplished by using the concepts of work and waste and the tools of variation, imagineering, and human relations.

Figure 14-1

The Right Way To Manage

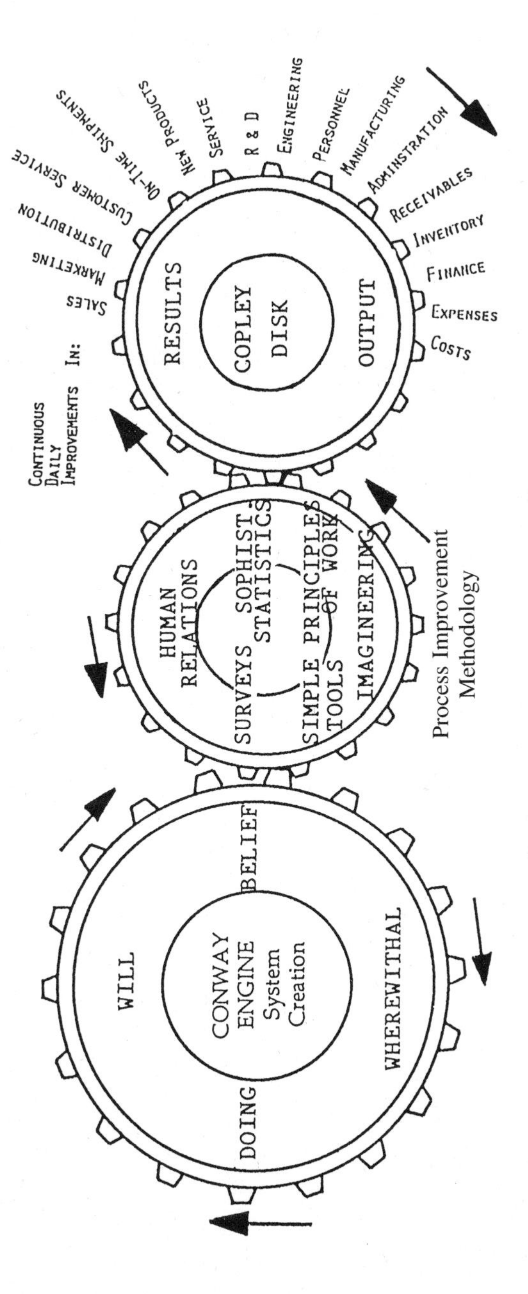

Productivity Through Quality Machine

Chapter 15
TAKING ACTION

The process of creating the system is different for each organization, depending especially on the organization's size and existing culture. A very large organization needs to plan how to choose pilot divisions or locations and how to phase in the process throughout all areas. Centralized, authoritarian organizations need to plan changes in their culture and strive in both words and deeds to convince people that the organization now wants their input and will listen.

This chapter discusses our "Quality Planning Process" that can be successfully modified to fit any organization that has leaders who have, or can develop, a strong commitment to make the necessary changes. A schematic of this system is shown in Figure 15-1.

COMMITTING TO CHANGE

For the process even to begin, a leader or leaders in the organization need a strong desire to change the status quo and the energy to influence others and commit resources to the effort. In order to take action, they must have the will, belief, and wherewithal described in the previous chapter. Eventually a critical mass of leaders in the organization must embrace this desire to change, believing that a shift to the new way of management is possible, practical, and capable of working miracles. Otherwise the will to carry it off will not be strong enough. This belief most often results from education and perhaps a leap of faith based on seeing what many companies in Japan and some in the West have accomplished using the new way.

Figure 15-1 Quality Planning Process

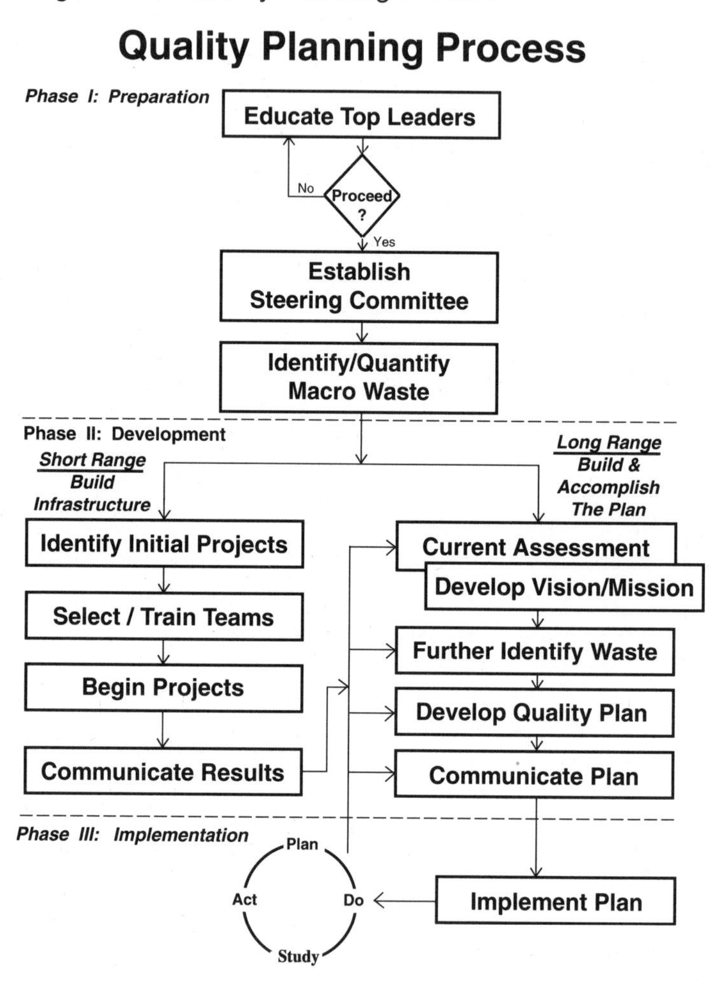

At some point, the leaders should make a commitment to start the shift to the new way. This commitment is not a casual or easy one. Nor can it be half-hearted. This process requires major changes and a major commitment. Sometimes in the past, a leader in an organization has started this process with the idea, "We can give it a small trial and see if anything comes of it." This approach is generally doomed to failure. Only a serious commitment will generate the momentum to overcome the barriers and resistance to change generated by such a management transformation.

Other leaders may be unwilling to make a commitment because of the risk of failure. Such fears can be self-fulfilling; failure is most likely to result from top management's lack of understanding and involvement. Education is so important because it gives those people critical to implementing the change an understanding of its power. In my experience the top leader and about 75% of the people reporting to the leader need to be committed to the change to sustain the momentum through the three phases of the quality planning process. These leaders must believe in the need for change, understand the concepts and tools of the new management system, and accept their roles as active leaders of the change.

THE QUALITY PLANNING PROCESS

Phase I: Preparation—Establishing a Steering Committee

The steering committee should be a team of the top leaders who will guide the transition of the company—or its division, plant, department, or business unit—into the new management system. Simply forming such a committee sends a message to the whole organization about the importance of the process. The committee should include the senior management team—the top five to ten people who run the business unit. The top quality person, if not already part of the top management team, should also be included. If union relations permit, consideration should be given to having a union representative. The committee may also include other key people who have special knowledge or who represent important constituencies.

The committee should be given a distinctive name to indicate the importance and special nature of the group. Organizations have used names like Quality Steering Committee, Quality Council, Quality Steering Team, and Continuous Improvement Support Group.

The head of the committee should be as highly placed within the organization as possible, to demonstrate the organization's commitment and set an example of active involvement. At first the committee should meet at least once a week, though the frequency can be reduced after it completes the initial planning. One of its initial tasks will be to define its purpose, something like, "Provide the ongoing direction and leadership to plan, implement, and maintain the Continuous Improvement Effort."

The committee will perform some of the planning activities itself and delegate other tasks to individuals or sub-groups. Typical responsibilities are to:

- Identify and quantify major areas of waste (macro waste). The steering committee must understand the location and extent of waste in the organization in order to realize the size of the opportunities, create a sense of urgency, and set priorities for major projects. One or more subcommittees or search teams can be assigned to develop data on the four areas of waste—material, capital, people's time and talent, and lost opportunities. This step of targeting the waste occurs during Phase I of the process. The following responsibilities will be accomplished during Phases II and III.

- Make a baseline assessment of the organization's present position. This should include the internal strengths and weaknesses, external threats and opportunities, and customers' needs and degree of satisfaction. If possible, the assessment should measure present levels of quality, customer satisfaction, productivity, and employee attitudes. These four areas should show dramatic improvement as the continuous improvement effort takes effect.

- Develop an implementation plan, including methods for measuring changes from the established baseline.

- Provide for the necessary education and training.
- Integrate the continuous improvement effort with the business plan.
- Encourage people working in the new system and publicize successes.
- Align the motivation/reward system with the continuous improvement effort.
- Provide necessary resources like a coordinator to represent the steering committee between meetings; one or more facilitators thoroughly trained in the tools of variation; and, for a large organization, an expert in statistical process control.
- Serve as active leaders and coaches for individual projects.
- Select major management-directed projects.
- Develop a mission and vision for the organization. This task will be discussed in detail later in this chapter.

Phase II: Development—Projects for Improvement

Phase II consists of two parallel activities: carrying out some projects for improvement and doing the detailed planning for the continuous improvement process.

The initial projects should be modest in scope and complexity. Their purpose is to begin the training process and generate some initial successes. It is very important that the first projects be successful. Everyone in the organization will be watching. Successes will ensure that those watching will want to be involved in similar projects. Therefore, both the projects and the participants should be chosen with care. The team members should be both competent at the work they're asked to do and enthusiastic about the new approach, so that they will serve as spokespeople and endorsers for the new process. The organization should be quick to provide the needed resources for training, guidance, and support.

These initial trials will provide a sense of how much staff support the projects need. The leaders, trainers, and facilitators should generally participate on a part-time basis while retaining their normal positions. People doing the organization's normal work every day understand how to be effective in the organization and will be accepted by their peers. Top leaders should support these people and treat them with respect to ensure that the rest of the organization perceives their special duties as a desirable assignment. As the process expands, leaders, trainers, and facilitators should be spread throughout the organization so the talent is ready in all areas to pursue continuous improvement. Always remember the best trainers and facilitators are the line managers.

The teams should receive their training in the use of the tools of variation just prior to starting their project(s). This "just-in-time" training will then be fresh in their minds, and they can enhance their learning by applying the tools to practical situations.

It is important to start some projects even before the planning is completed. Action tends to generate momentum which is otherwise absent during a lengthy planning, education, and training period. These projects can create excitement and a desire by those not yet working the new way to receive the necessary education and training and to get started.

Publicizing the successes of the early projects also builds the excitement and momentum. The top leaders' recognition of successful project teams sends a strong message to everyone that continuous improvement efforts will be rewarded.

This work continues to expand. More projects start, more people get involved, and the scope of the projects broadens while the Phase II planning effort is underway. This activity builds an infrastructure for working the new way, as people learn enough to be able to plan.

Phase II: Development—Planning

The steering committee must undertake, or direct, the detailed planning of its various responsibilities.

Education and training. Presumably, by this point the steering committee members themselves have received some education and training. Other employees should be educated fairly quickly to underline the organization's commitment and to ensure that people understand what the organization is doing and why.

Most organizations use a combination of seminars by outsiders or insiders, videotapes, reading, and coaching. Plans should be made for some education at each level. First the organization should continue the education of all the top leaders, then include mid-level managers and union or work force leaders. A more condensed version should be offered for all first level supervisors, and eventually all employees should be trained. The employees must understand that they are the experts and will be expected to contribute their knowledge and experience to the continuous improvement effort. This process cannot be "installed" by an outside consultant. It is a shift to a new way of working that must gradually be embraced by everyone in the organization.

Where possible, the training in how to use the tools of variation in project work should be deferred until just before a project team begins work. Individual internal or external trainers/facilitators should be developed to work part-time or full-time. Ideally the project team should be trained as a group. The classes should be fairly small, as should the team (no more than seven to nine).

Human relations planning. The organization must also educate everyone about the human relations aspects of the new way of working. For the changes to be successful, people must perceive that it is in their interest to cooperate. They should welcome the changes that give them more input into, and control of, the work process. What concerns people most is how the drive for continuous improvement will affect their employment security. In the long run the only true employment security comes from having top-quality, low-cost products and services that external customers want. Continuous improvement will require a smaller number of people to do the same amount of work.

Management must decide how it will handle that situation. Companies working on continuous improvement often begin to do more business and therefore need more people. But managers working in the new way prefer to talk about employment security rather than job security because, as we have seen, getting rid of the waste in one area often means that people are freed to move to another area where they can be more useful. While no management can guarantee against job losses for any reason, organizations can often pledge that no one will be laid off because of the gains of the continuous improvement program. This commitment may spur workers to find and eliminate waste that they might otherwise guard for fear that the company might cut their jobs with the waste. The best answer is always more business. If management is concerned about making a public pledge they should at least show by their actions that they are doing everything possible to provide employment security to those people who adopt and support the new system.

At the top of the list of other human relations considerations are personnel policies, which should be reviewed to determine which ones are inconsistent with the new way of managing. Union leaders or leaders from the workforce should be consulted about plans made to change these policies. Most organizations will require five to ten years to make many of these changes. The cooperation of the work force is vital to the process of continuous improvement.

The organization should prepare a specific plan for communicating with all employees the purpose of the process, its status, and the progress being made. Regular communications in the form of a newsletter or a letter from the steering committee can be effective. Even more important is what the committee members say and do about the program daily. They must serve as missionaries by discussing it frequently and showing their enthusiastic leadership and active visible involvement in the process.

Part of the communication plan should be a system of recognizing individuals and teams who embrace the process. Management should modify the normal recognition/reward system of the organization to incorporate and emphasize work done in the new system. One effective form of recognition is to have project teams present their accomplishments to groups including top management. This serves the dual purpose of recognition and training. People are motivated by examples of success and learn from them.

Phase II: Development—Defining Mission and Vision

A statement of mission defines an organization. It gives the core reason for the organization's existence. Although most people assume they know what their business is, I have been amazed

Figure 15-2 Mission/Vision Process

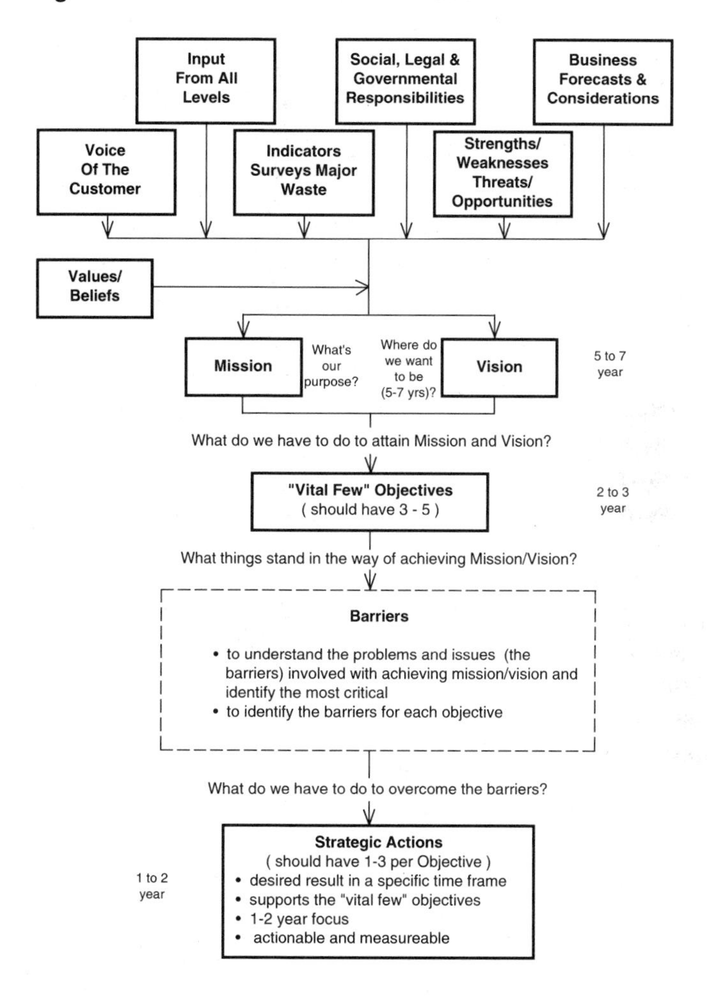

work installation by at least 10% per year over the next three years, while improving performance."

These objectives identify the things that need to receive high priority attention. They provide the keys to progress toward the vision and help identify the barriers that have to be surmounted. The steering committee should study each of the vital objectives to determine the specific barriers to be overcome and the specific strategic actions needed to accomplish each objective. The computer networking company might identify as a barrier to its international market share objective "Too many of the strong European distributors for our services are already committed to our competitors." A project or strategic action to overcome that barrier might be "Achieve at least a 25% market share in Europe within three years by a two-pronged effort: 1) Develop a program to entice strong distributors to switch to our product; and 2) Investigate the acquisition of distribution companies in key countries."

This strategic action is specific enough for the steering committee to assign to a team as a project, one of perhaps six to twelve management-directed projects given priority by the company to achieve its vision. The steering committee should insure that these project teams have the necessary resources and the total support of management. A regular reporting system should be established to keep management informed of progress and problems.

Phase II: Development and Communication of the Continuous Improvement Plan

In parallel to the development of the vision and mission, a detailed implementation plan should be created. The steering committee may assign this task to a subcommittee or a different team altogether, but it needs to review and agree to the final plans. The team should detail the plans for communication, education, training, project team assignment, and project review and evaluation.

One or more members of this planning team should be made responsible for coordinating the plan implementation on a full-time or part-time basis. This person(s) should be responsible for providing the training and other support resources for the project teams. He or she should also coordinate the project reporting system and serve as liaison between the project teams and the steering committee. Someone should also be assigned to execute the communication plan. He or she should insure that everyone in the organization is kept up to date on training schedules, project progress, and the plans of the steering committee. Strong information flow is needed to help create momentum for the new system.

The project reporting system should be simple and flexible, so that it is not just an added bureaucratic burden. Since every project is different, too much standardization of reports would be a constraint. Each team should, however, provide charts of its key operating variables and a short written description of its progress and problems. Managers should expect these charts to be kept current and should informally review them frequently. An updated chart is a sign of an active project. An outdated one is a sign that nothing is happening. Managers should also make sure that everyone hears about successes.

Phase III: Implementation

There is some overlap between Phases II and III since pilot teams are set up and work starts during Phase II. Once project work begins, participants learn by doing. Communication among the teams and between teams and the steering committee is especially important at the beginning. Problems and successes both need to be widely discussed, since discussion accelerates the learning process.

How quickly to form new teams is an important consideration. If the steering committee has done its job, there will be more requests for new project teams than the training can accommodate. This is a good sign, but proliferation of projects must be restricted at first. Teams formed without first being trained to use the tools of variation will flounder. Also, since people learn by early mistakes, too many teams too soon will mean repetition of a lot of mistakes. Teams that don't do well and get discouraged hurt the process throughout the organization.

Once an organization has had several successes and the training is widespread, people will form spontaneous teams to attack specific areas of concern to them. While team formation is a healthy sign, the organization must keep a balance between the support resources available and the number of teams in operation. Although some teams may be successful on their own, many will not be without some guidance from someone further down the learning curve. Keeping this balance must be the job of the steering committee, working through the managers of the organization.

During implementation, the degree of success and pace of progress is highly dependent on management leadership. The amount of time members of the steering committee spend and their dedication is key to involving everyone in the process and to incorporating the drive for continuous improvement into the culture of the organization.

Chapter 16

QUESTIONS THAT SPUR PROCESS IMPROVEMENT

"They don't know what to look for. They don't even know what questions to ask."[11] When Dr. Deming made this statement about U.S. executives visiting Japanese companies, people reacted defensively. But Dr. Deming was not implying that asking appropriate questions is easy. Asking the right questions is difficult and must be preceded by a lot of work and thought to understand thoroughly what you are questioning.

It is a critical skill for managing in the new way. Continuous improvement requires change, and much of that change is initiated by asking the right questions about the way things are being done now. The right questions increase the alternatives for attacking problems and opportunities. They force people to examine new directions and get the creative juices flowing. The right question can sometimes make a dramatic improvement in the way an organization works. For these reasons, it is important to examine the process of questioning the status quo.

To ask the right questions, first you need a thorough knowledge of the work and work processes. As Edison said, genius is 99% perspiration and only 1% inspiration. To ask a perceptive question, you must first perceive what is happening in the work process. Walking around and observing is fine but often not enough. You must be willing to put forth the effort to learn in some detail what is happening to be able to see how things might be done differently.

The second requirement is perception, understanding or insight that goes beyond merely seeing something. Some people are naturally more perceptive than others, but they have become that way by being observant, noticing differences, and continually asking themselves what caused those differences and how things could be changed.

Perceptive people also are curious and dissatisfied. At work, they are curious about how the work is done and where the waste occurs. They find some waste and are suspicious that there is much more. Then they gather facts. They look for variation, for differences in the way things happen. They interpret facts about the differences by using statistical tools. Often they perceive as important facts that others dismiss. They may find that one salesperson's calling pattern, one customer's packaging, one order clerk's work process, is better than the others.

To turn perception into action, you must be dissatisfied. Unless you are dissatisfied with the present, you will never ask the right questions and make the required changes. Often people use budgets or other goals and say, "I'm doing fine; I'm 20% over my projection." With that attitude they will remain on a plateau and not improve continuously. People must have a sense of urgency, because resistance to change is endemic in most organizations.

Besides this sense of urgency, anyone trying to change an organization must have a sensitivity to the needs of the people in the organization. Overcoming resistance to change is a topic that could fill another book. In essence, people must perceive that changing is in their own best interests because, for instance, it will provide them with a more meaningful job or more employment security as it makes their organization stronger.

The most successful questioners learn to pick an important area and then focus on it intently, look at all its aspects over a period of time, and question how it could be done differently. They don't accept the obvious way to structure a work process. They begin, in their imagination, with a clean slate and try to see how it should be done if the operation were just starting.

Several types of questions are valuable in opening people's minds and encouraging continuous improvement. One type is asking why something is happening five times. For example: 1) Why are the pumps out of service so often? Because the bearings overheat and bind. 2) Why? Because they are not lubricated properly. 3) Why? Because proper maintenance procedures are not followed. 4) Why? Because maintenance procedures are outdated and apply to a previous model of pump. 5) Why? No one has the assigned responsibility to insure that maintenance manuals are updated.

This type of questioning helps get to the root source of a problem so that when it is fixed, it will stay fixed. Sometimes a sixth or seventh "way" may be useful. A sixth why in the above example might bring a reply: "Because company culture, policies and procedures are loose about assigning responsibility." The more we ask why, the more chance we have of reaching the basic causes of a problem.

The most powerful type of question is the all-encompassing imagineering question: **"What would this process be like if everything were perfect?"** This question is powerful because it encourages ideas for change without creating a defensive posture, since everyone recognizes that nothing is ever perfect.

To answer this question effectively, you must have a lot of facts. With those facts you can examine the variation:

What are the differences we can identify?

From perfection?
From competitors?
In vendors?
In the way different people work?
In the way different pieces of equipment work?
From one period (minute, hour, shift, day, week, month) to another?

Where does all the money go? In Chapter 8 I described how I got Jim Copley's attention by asking him what happened to the $55,000,000 of his sales which were not profit. The initial answers to such a question are usually too broad-brush to be of any value. But by starting with the broad brush and then working down into more and more detail, you can begin to see, in a new way, just where the opportunities for improvement lie.

Where is the waste? As Jim Copley went into more detail, he found it easier to identify the sources of waste, to gather data about them, and to find ways to eliminate them.

How can we level the work and/or vary the work force? One of the greatest sources of waste is the mismatch between the amount of value-added work to be done and the resources available to perform that work. Work does not flow evenly, particularly in non-direct areas, and consequently employees waste time waiting for work or filling in with make-work activities. Creative scheduling to level the work and shifting of resources to match work flow needs can eliminate much of this waste.

What does the customer really want? Meeting customer needs is the key to viability in any business. In the mail order photofinishing business, distributing envelopes promoting a low introductory price is a major activity and expense. The promotion is intended to attract customers with a low price, and the company hopes that its high quality and service will make the customers return with more business at a normal price. The problem with this strategy is that customers develop little sense of loyalty to one company. They want the lowest possible price all the time, so they will just look for any promotional envelope, regardless of company.

When we faced this problem at Nashua, I asked why we didn't try meeting the customer's needs by charging the same low price for repeat orders as we offered for introductory ones. Although it was "obvious" that this would be too costly, we arranged a test of the concept. The resulting increase in customer loyalty was phenomenal. Reduced marketing costs made up for much of the lower prices, and sales began to mushroom. We also developed cost reduction programs by improving processes in all areas of the business. This proved to be a major turning point: a business that was losing money grew to be extremely profitable and became the world's largest direct mail photofinisher, a truly world class business. If you can find a way to meet customers' needs that none of your competitors has discovered, the rewards will be almost automatic.

Is there anything we could sell more of if we had the capacity? Most organizations cannot at times meet the demand for a product or service because of limited capacity. Such organizations should ask themselves, "What are the bottlenecks and how do we eliminate them?" A creative questioning of the bottlenecks can often lead to relatively inexpensive ways to reduce or eliminate them. Posing some key questions to a major world oil company led to its optimizing production of a particular light crude from its wells. The premium price on this crude meant these efforts produced hundreds of millions of extra dollars annually.

What do the employees really want? This question can be almost as powerful as asking what the customers really want. People must perceive that it is in their interest to be a part of the team working on continuous improvement. Organizations must be able to address their needs for meaningful work and reasonable employment security.

What is the "pace" of the organization? Does management set a good example and have a sense of urgency? Do people feel they want to put in a good day's work to help the team, or are some of them doing just enough to get by?

Do all of the people have full-time, meaningful work? A "no" answer is the most common cause of a slack pace. It is management's responsibility to organize operations so that the answer is "yes."

How many organizational levels are there? Could there be fewer? Usually, the fewer the levels between the chief executive and the lowest paid workers, the more effective the organization. Companies with under 500 employees should question why they need more than four levels, and even the largest company should be able to operate with no more than eight.

How can we expand the business so that we can utilize the people as we find and eliminate waste in the work? This is another management responsibility. Normal attrition may be part of the answer. In any case, some answer must be found to address the problem of employment security or you will get tremendous resistance from people who are asked to help eliminate waste of people's time.

How do our customers feel about the company? If you do not know, it is important to find out. Usually the best source of more business is existing customers.

What would happen to our sales if our quality were better than our competitors' and our costs 20% lower? The objective of this book is to enable you to reach this enviable position. It leads to the world's best marketing program because you are now in position to reduce prices, add other value, develop new products, make further improvements in quality, and lower costs.

How can we improve process capability? What does the variation in the process tell us?

Are our specifications too tight or too loose?

How can we make our suppliers part of our team? An organization can develop a "win-win" situation by working closely with some of its suppliers to help them provide the products or services the organization needs at the right price.

Are we the low-cost producer? Can we be? How?

Are people working on the right things?
How are priorities set?
Are the right things being measured?
Is action being taken on the facts and data?

Are there barriers between the various functions in our organization? Does each functional area, such as sales or manufacturing, focus only on making its own performance look good, or is every area working for the overall good of the company?

Do our accounting and performance measurement systems drive people to do the wrong thing? A purchasing manager measured solely on reducing the cost of purchased materials will tend to choose the lowest price, regardless of problems the material may cause in production. A production superintendent measured on unit costs may favor long production runs which result in excess inventory. In general, if performance measurements for each functional area are not carefully designed, they tend to concentrate effort on optimizing the particular function, often at the expense of the overall organization.

What currently impossible goal would fundamentally change our business or our operations if we could accomplish it in the future? Often organizations develop blind spots or fixed ideas about what they can accomplish. They may overlook such things as the possibility of developing a new product, of reducing costs or improving quality in a major way, of penetrating a new market, or of increasing market share substantially. Imagining the "impossible" to be possible can help people accept a paradigm shift in the way the organization operates, opening the way for vast changes and improvements in the organization.

What possibly could happen that would damage our business fundamentally? This is the negative side of the previous question. Thinking about such possibilities and how to prevent them may help to forestall disaster.

What actions would another company take if it acquired our organization? This is another question that opens minds to the possibility of major changes to the organization.

In what areas do some of our competitors perform better than we do? Quality? Costs? Sales or value-added per payroll dollar? Customer acceptance? Other areas? An organization that is not already the best should compare itself to the best to provide a realistic target or benchmark for improvement.

What are the reasons we lost orders to competitors? Find the answers with a lost order report or other means. This is another way of discovering areas that need improvement.

Do we have a clearly defined mission and vision that have been communicated to all employees? People need to understand what the organization is trying to do if they are to be effective in helping to reach those goals.

Do we have a quality plan for our department, group, business unit? This is the plan to move from where we are to where we want to be one year, three years from now.

Why can't we do it? Just go do it! A sense of urgency is often needed, especially if the desired improvement seems difficult or impossible. I was often told we had done everything possible to reduce the cost of mailing pictures back to the customer in our photofinishing business. Anything other than first class mail delayed delivery unacceptably, except within a few hundred miles of our plant. We couldn't negotiate with the U.S. Post Office for lower first class rates. Even though I did not know whether it was possible, I insisted that we study the mail distribution system and find a way to reduce postage costs substantially. We found we could mail first-class bulk express overnight to postal sectional centers and then send individual orders third class from there without increasing delivery time. We reduced our postage costs by 35%, and since they were such a major expense, our total production and distribution costs dropped by 7%.

Continuous improvement requires continual change. Perceptive questions are catalysts for that change.

Chapter 17
The Quality Secret

THE QUALITY SECRET

Improving quality creates the *opportunity* to lower costs and raise productivity dramatically. To achieve the full benefits of quality improvement, management must capture these potential gains. The Right Way to Manage helps you to do this while at the same time pleasing your external customers. Many organizations today are claiming quality while they are totally non-competitive in costs.

Quality is process driven. To improve the quality of a product or service you must improve the process(es) that produce and support that product or service. Conventional wisdom requires that you pick the "most important" or "critical" processes and work to improve them. The problem with that approach is that you do not know if you are working on the right things to get rid of the waste and prevent its return. It makes more sense to find the waste first and then determine the right processes to improve. Measure the process and the waste. In The Right Way to Manage, the focus is on finding the waste, working on the "most important" causes first, and permanently getting rid of those causes of waste, one by one, through process improvement. "Most important" includes those that require the least capital investment or are easiest or quickest to fix, largest, or most important to external customers. Measure the process improvements and the waste improvements to make sure you are working on the right things in the right processes.

Waste results from what is being worked on and the way the work is being done, whether by machines, chemicals, computers, or people. If you eliminate the waste from a process, most of the work remaining will be value-added work. Probably the most important process in any organization is the process by which everyone at every level uses the concept of value-added work to decide what to work on and to improve that work constantly. Toyota is the world leader in quality because its management focuses constantly on the value-added concept and on finding waste, eliminating it, and keeping it gone. These focuses are so important to the people at Toyota that the average employee makes over 50 suggestions per year, 96% of which are implemented.

The QUALITY SECRET is that the elimination of the waste in a process automatically improves the quality coming from that process. By reducing the problems and variation in a process, stabilizing it, and producing what the customer wants, you can achieve high consistent quality at low cost. The Right Way to Manage requires that you find out what the customer wants and needs and work continuously to improve the processes that provide and support that product or service. Many quality systems advocate this improvement but fail to provide a system that insures you work on the right things and capture potential gains. The right way to achieve process improvement is through the core activity of identifying, quantifying, and eliminating waste, including waste from non-value-added work. This activity, when done on a continuous basis by people educated in the use of the tools of process improvement, will bring higher and higher quality at lower and lower costs and can result in a world-class operation. That is the quality secret.

Now that you have finished this book, you must make a decision. Do you believe in what you have read to the degree that you are willing to accept wholeheartedly this new model and take action based on it? There are no halfway measures. Many people have "experimented" with this model in a small way, only to have it slide into oblivion under the press of everyday problems. The changes in thinking are so large that a major commitment and vigorous, persistent action are necessary to adopt this new system of management successfully. You can be skeptical of a few things at the start, but you must be ready to learn, try out new ways, and change as you move forward. Although this system is difficult to execute, its rewards are dramatic. In this increasingly competitive world, consider the alternative.

Appendix A
Concepts and Principles

To complement the introduction's overview of the book's major ideas, this appendix summarizes the principles that drive the "Right Way to Manage." I encourage you to review these periodically to insure that you are making them part of the way you think, talk, work, and act.

1. The core activity of the "Right Way to Manage" is to *improve continuously all work processes by identifying, quantifying, and eliminating waste.*

This approach differs from conventional management primarily because

a. Continuous improvement means continuous change, a continuous drive to do it better. Conventional management works either to maintain the status quo or to set finite, limited goals for improvement.

b. Conventional management looks for waste only in its most obvious forms. The new system looks everywhere for waste and scrutinizes every activity and process that does not add value from the external customers' viewpoint. It provides a methodical way to identify waste with the objective of continuously working to eliminate it.

c. In the new system, process improvement increases both quality and productivity, correcting problems when they first happen, thus reducing waste and rework. The old system often sacrifices quality for productivity, or vice versa.

d. Continuous improvement includes not only improving the work and work processes but also working on the right things at every level, every area of the organization.

2. *The four forms of waste are waste of material, waste of capital, waste of people's time, energy and brainpower, and waste from lost sales and opportunities.* Of these four the last two are usually the most important sources of waste, as organizations make inadequate use of people and fail to recognize and exploit opportunities.

WORK

3. *All the forms of waste ultimately come from work and the processes used to accomplish work.* In other words, waste results from what organizations work on and how they work. All forms of work can produce waste—work done by such things as machines, chemical processes, and energy, as well as that done by people.

4. To eliminate waste requires a detailed knowledge of the work itself, the kind of knowledge that only the people operating the processes usually command. Therefore *organizations must enlist the help of the people operating the processes to identify, quantify, and eliminate waste.*

5. *If you eliminate the waste from your work, almost all the remaining work will have value for your customer and thus be real value-added work.* Organizations must strive constantly to improve the percentage of activities that add value while improving the way they do value-added work. This concept is at the heart of continuous improvement. Always remember that it is crucial to decide what to work on. If that decision is wrong, so is all else.

6. *Organizations should staff and schedule operations so that employees have full-time, value-added work available at all times.* Although this is difficult to accomplish, not doing it is one

of the largest causes of waste. If people don't have value-added work available, they find something else to do to fill the time.

7. *A powerful tool for discovering waste is to compare mentally the way things are now and the "imagineered" way they would be if everything were perfect.* Simply fantasizing perfection is not very useful. Successful imagineering depends on gathering facts, creatively questioning the work process, and understanding in detail how the process works.

CUSTOMERS and SUPPLIERS

8. *The underlying mission of any viable organization must be to please customers.* Pleasing customers requires communicating to understand their wants and needs. Most successful organizations also anticipate what the customer will want in the future and make continuous improvements with the customer in mind. The ultimate customer to please is the external customer who uses the organization's product or service.

9. *The efficient production of quality goods and services for customers requires that suppliers also work in the new way.* The organization must communicate well with its suppliers and educate them about the organization's requirements. Because such communication and education takes time, the organization should minimize the number of suppliers it uses. Ideally, the organization will have just one supplier for each major item it purchases, and that supplier will work so closely with the organization that it will become part of the organization's team, supporting the vision of the organization.

10. *The best possible marketing tool is to provide a high quality product or service with low cost.* This goal simplifies the overall marketing plan. High quality at low cost is also the only real employment security.

HUMAN RELATIONS

11. *"Treat people as you would like to be treated"* is the golden rule that encourages people to work in the new system. Leaders

working this way, using employees' talents to advance continuous improvement, help people raise their self-esteem.

12. *The process or the system, and not the individual, is responsible for 90% or more of all problems, waste, and lost opportunities.* Rather than blame the individual for process problems, the organization must enlist his or her help to identify and eliminate those problems that cause waste. The people operating a process know its problems better than anyone else and are thus in the best position to suggest improvements. But since they are working in the system, they need the help of managers working on the system.

13. *People in the organization should perceive that it is in their interest to share their knowledge of the best way to work.* No matter how complex or simple the work, if the organization wants continuous improvement, it must learn the best methods from the employees who are practicing those methods. And those employees must see that it is in their interest to become team members and teachers and share their knowledge.

14. *Work standards, numerical goals, and objectives often artificially limit quality and productivity.* If people fear NOT meeting objectives, they will try to establish objectives they are sure they can readily meet. In the new system the goal is continuous improvement, so work towards the goal is never finished.

15. *Developing a culture that makes people feel part of a team is the only effective way of fully utilizing people's time, energy, and brainpower.* This culture communicates the knowledge that once people work successfully in the new way they can help others do the same. The feeling of being a valued member of a team that supports the mission and vision of an organization generates motivation far greater than that provided by a paycheck alone.

VARIATION

16. *The statistical tools which identify, analyze, and communicate variation are key to finding and eliminating waste.* Studying the variation in a process shows where the waste occurs. Keeping charts of the fundamental operating variables measures

progress, warns of adverse changes, and focuses effort on the things that count.

17. *Control charts are invaluable in achieving, maintaining, and improving a stable process.*

a. Identifying and eliminating special causes stabilizes a process, leaving only common causes of variation which are inherent in the way the process operates.

b. Once a process is stable, work on common causes can narrow the process's range of variation. There will always be some variation in a process, but work on continuous reduction of variation will yield continuous improvement.

c. Control charts are essential to measure accurately the effects of experiments or trial changes. Separating the random variation in a process from that caused by a deliberate change makes clear the effect of the change itself.

18. *Charts provide the means for measuring continuous improvement.* The measurement of progress helps to focus effort on the things that count and indicates when a new or different approach may be necessary. Everyone needs to maintain a Pareto mentality to be sure the organization is measuring and working on the things that count.

IMPLEMENTATION

19. *To work as a team, people need the kind of common purpose provided by a mission statement that describes the basic function of the organization and a vision statement that sets forth the organization's ambitious but realistic goals.* These statements are basic to the quality plan, providing the focus, direction, and constancy of purpose to keep everyone working together. With the vision in mind, the organization's leaders can develop more specific objectives and projects to achieve those objectives.

20. *A small number of major management-directed projects should receive priority of resources and management attention* once the organization has built the infrastructure of working the new way. To avoid trying to do too much at once and to insure that

critical things are accomplished, management must identify those six to twelve projects most essential to achieving the vision.

21. *Leaders in the new system must be leaders of change.* They take risks and set an example for continuous improvement by being active, visible participants in making changes. By their actions and words they constantly promote working in the new way to support the vision of the organization. Managers need to be leaders, but leaders need not be managers, and can come from any level of the organization.

22. *Converting to the new system requires the will to do it, the belief that it can be done, the wherewithal to accomplish it in the form of education and training, and action.* These four items make up the engine of continuous improvement. Leadership provides the energy to drive this engine and to make continuous improvement everyone's job—forever.

Once you begin working in the new way, studying the variation, using the charts, changing the culture, and getting excited by the results of your work, most of these principles will become second nature to you, and you will probably develop your own principles for applying The Right Way to Manage to your own organization. After a few months of ferreting out and eliminating waste, you may not need to look back at the "rules" any more than a good writer needs to review the rules of grammar. But like such a writer, you will be hunting down waste, redundancy, and flabbiness, and coming up with creative and efficient ways to get rid of it, almost automatically. The system will be in your blood. And your organization—and with any luck, eventually the entire nation—will benefit.

NOTES

1. Copyright 1983 by Conway Quality, Inc. All rights reserved.

2. W. A. Shewhart, Economic Control of Quality of Manufactured Product (New York: Van Nostrand, 1931), republished in 1980 by American Society for Quality Control, 161 W. Wisconsin Avenue, Milwaukee, WI 53203.

3. Micheline Maynard, "Overtaking the Big Three on Their Turf," USA Today, September 11, 1990, 1b–2b.

4. The Economist, February 23, 1980.

5. Peter Drucker, "Low Wages No Longer Give Competitive Edge," The Wall Street Journal, April 16, 1988, 30.

6. Leonard Murray, OBE

7. Dr. Lloyd S. Nelson, internal memo, Nashua Corporation

8. Julius Alexander and Connie O'Daniels, "Center Wing Feedback System," Lockheed Aeronautical Systems Company, 1991.

9. Kaoru Ishikawa, Guide to Quality Control (Asian Productivity Organization, 1976). Available from UNIPUB, P.O. Box 433, Murray Hill Station, New York, NY 10157, Tel. 800-521-8110.

10. Dr. Lloyd S. Nelson, teaching notes

11. Dr. W. Edwards Deming, "If Japan Can, Why Can't We?," NBC white paper, June 1981.

12. E.F. Wells, "Imagineering," American Airlines Magazine, date unknown.

13. William Thomson, Lord Kelvin

14. Taiichi Ohno, Workplace Management (Cambridge: Productivity Press, 1988).

15. R. W. Hall, Zero Inventories (Homewood, IL.: Dow Jones Irwin, 1983).

16. Daniel J. Boorstin, The Discoverers (New York: Random House, 1983), 86.

17. Robert H. Hayes and Steven C. Wheelwright, Restoring Our Competitive Edge (New York: John Wiley and Sons, 1984), 8.

18. Niccolo Machiavelli, The Prince, volume 23 of Great Books of the Western World (Chicago: Encyclopedia Britannica, 1952), 9. Originally published in 1513.

RESOURCES

1. **Out of the Crisis**
   W. Edwards Deming, Massachusetts Institute of Technology, Center for Advanced Engineering Study, Cambridge, MA 02139. 1986. Telephone 617-253-7444.
2. **Quality Planning and Analysis**
   Juran/Gryna. Second Edition. McGraw-Hill Book Company, 1980. Telephone 212-512-2000.
3. **Guide to Quality Control**
   Ishikawa. UNIPUB. Box 433, Murray Hill Station, NY, NY 10157. Telephone 800-521-8110.
4. **Employment Security in a Free Economy**
   Jerome Rostow. Pergamon Press. Elmsford, NY. Telephone 914-592-7700.
5. **Competitive Advantage**
   Michael E. Porter. Free Press. 866 3rd Avenue, NY, NY 10022. Telephone 212-702-2000.
6. **Non-Stock Production**
   Shigeo Shingo. Productivity Press, 1988. Cambridge, MA 02140. Telephone 617-497-5146.
7. **World Class Manufacturing**
   Richard J. Schonberger. Free Press, 1986. 866 Third Avenue, NY, NY 10022. Telephone 212-702-2000.
8. **Japanese Manufacturing Techniques**
   Richard J. Schonberger. Free Press, 1982. 866 Third Avenue, NY, NY 10022. Telephone 212-702-2000.
9. **Toyota Production System**
   Taiichi Ohno. Productivity Press, 1988. Cambridge, MA 02140. Telephone 617-497-5146.
10. **Zero Inventories**
    R.W. Hall. Dow Jones-Irwin, 1983. Homewood, IL 60430.

11. **Quality Progress Magazine**
American Society for Quality Control, 310 West Wisconsin Avenue, Milwaukee, WI 53203. Telephone 414-272-8575.

12. **The Goal**
Eliyahu M. Goldratt and Jeff Cox. North River Press, Inc. 1986. Box 309 Croton-on-Hudson, NY 10520. Telephone 914-941-7175.

13. **Just-In-Time for Today and Tomorrow**
Taiichi Ohno with Setsuo Mito. Productivity Press, Cambridge, MA 02140. Telephone 617-497-5146.

14. **The Deming Management Method**
Mary Walton. Putnam Publishing Group, 200 Madison Avenue, NY, NY 10016.

15. **The Deming Route to Quality and Productivity**
William W. Scherkenbach. Mercury Press/Fairchild Publications, Rockville, MD 20852. Telephone 301-770-6177.

16. **SPC Simplified**
Robert T. Amsden, Howard E. Butler, David M. Amsden. UNIPUB, 1986. Kraus International Publications, White Plains, NY.

17. **Future Perfect**
Stanley M. Davis. Addison-Wesley Publishing Company, Inc. 1987. Reading, MA.

18. **The Team Handbook**
Joiner Associates, Inc., 3800 Regent Street, Madison, WI 53705-0445. Telephone 608-238-8134.

19. **Productivity, Inc.—News Letter**
Productivity Press, Cambridge, MA 02140. Telephone 617-497-5146.

20. **Better Designs In Half the Time**
Bob King. GOAL/QPC, 13 Branch Street, Methuen, MA. 01844. Telephone 508-685-3900.

21. **How to Win Friends and Influence People**
Dale Carnegie. Simon & Schuster, Inc., 1230 Avenue of the Americas, New York, NY 10020.

22. **Workplace Management**
Taiichi Ohno. Productivity Press, 1988. Cambridge, MA 02140. Telephone 617-497-5146.

23. **The Just-In-Time Breakthrough**
Edward J. Hay. John Wiley & Sons, Inc. 605 3rd Avenue, New York, NY 10158-0012. Telephone 212-850-6418.

24. **The Memory Jogger & The Memory Jogger Plus**
Michael Brassard. GOAL/QPC, 13 Branch Street, Methuen, MA 01844. Telephone 508-685-3900.

25. **Video Courses**
Conway Quality, Inc., 15 Trafalgar Square, Nashua, NH 03063. Telephone 800-359-0099

What Is the New Management System? (75 mins.)
Creating the New Management System (90 mins.)
Four Forms of Waste (64 mins.)
Five Approaches to Identifying Waste (80 mins.)
Key Concepts of Variation (110 mins.)
Imagineering (70 mins.)
Eliminating Waste (54 mins.)
Principles of Work: An Overview (80 mins.)
Work Sampling (65 mins.)
Process Flow Charts/Work Improvement (65 mins.)
Human Relations for Continuous Improvement (85 mins.)
Leadership for Continuous Improvement (85 mins.)
Evaluating Progress in the New Management System (85 mins.)
Continuous Improvement at Shipley (58 mins.)
Quality Performance at Dow (62 mins.)
Bill Conway Live (75 mins.)

INDEX